T0135776

EXPLORATORY MODEL COMPARISON

Interactive Model Ensemble Selection And Management

Ralf Seger

Augsburger Schriften zur Mathematik, Physik und Informatik
Band 17

herausgegeben von:
Professor Dr. F. Pukelsheim
Professor Dr. W. Reif
Professor Dr. D. Vollhardt

Bibliografische Information der Deutschen Nationalbibliothek

Die Deutsche Nationalbibliothek verzeichnet diese Publikation in der
Deutschen Nationalbibliografie; detaillierte bibliografische Daten sind
im Internet über http://dnb.d-nb.de abrufbar.

ISBN 978-3-8325-2927-7
ISSN 1611-4256

Logos Verlag Berlin GmbH
Comeniushof, Gubener Str. 47,
10243 Berlin
Tel.: +49 030 42 85 10 90
Fax: +49 030 42 85 10 92
INTERNET: http://www.logos-verlag.de

Contents

All models are wrong, but some are useful.

George E.P. Box

Preface

This work is concerned with statistical models. Yet I start with a quotation that all models are wrong. Having spent so much time with models over recent years, I have came to believe this quotation is more true than ever before. Models are not meant to be right. Reality is not formed in such a way that any model can capture. Nevertheless models are very helpful to understand this reality we live in. By the way, this is one reason not to rely on just one best model for inference. Yet even today most practitioners shy away from the extra effort involved in working with more than one model, although they know well the advantages of using model ensembles.

I do not want to dive too deeply into philosophy, neither in this preface nor in the work itself. Models and software about models kept me busy for more years than I ever anticipated. Still the fascination is not lost and I feel I have to give credit to my advisor Antony Unwin for his patience in guiding me thus far. He introduced me to statistical models in his lectures and later opened the world of ensemble models for me. Without his endurance and encouragements I would never have completed this. And *this* does not merely cover this book but also the software MORET, I've been writing. In the course of time I learned many things, not only about models but lots of techniques how to create software. Since software is usually too old when it's released for the first time I started a follow up project so that others may benefit even after this prototype is no longer actively developed.

http://r-forge.r-project.org/projects/joris/ is the successor and meant to be a basic platform for R and java based software. The focus is to provide a database interface and means to store and retrieve statistical models from this platform. MORET will be refactored to be a graphical front end for this project in the future. Any request for improvements or features are highly appreciated.

My experience, both errors and insights collected over these years, may assist anyone interested in working with statistical models.

Augsburg, April 30, Ralf Seger

Chapter 1

Introduction

1.1 What Is A Model ?

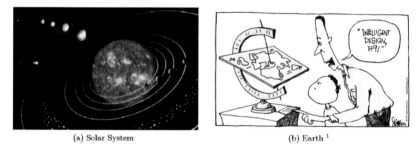

(a) Solar System   (b) Earth [1]

Figure 1.1: Models

A model is an abstraction of reality. By simplifying the complex world we live in a model allows us to grasp difficult concepts and comprehend the underlying structures. Our universe, for example, consisting of countless entities from the smallest particle to galaxies, is very intricate. The Earth, the world we live on, consists of such a huge variety and number of objects that the interaction of all participating objects is far too complex to grasp. Looking at this earth from the outer space perspective we perceive a roughly spheroid shape. A globe is a model for our planet, but not a very helpful one since a large enough globe to locate houses or streets is far too bulky to transport. The model of choice to locate objects on earth is a map. A map does not care about the spheroid[2] shape. Reducing accuracy in this model serves the purpose better. The story of maps and models of our planet is a long one. Today we can measure the components

[2]Different transformations care about flattening the spheroid shape. See for instance the Universal Transverse Mercator coordinate system

this globe is made of. We can trace the orbits of other celestial entities with large telescopes and calculate their characteristics accurately with the help of modern computers. To produce accurate maps modern techniques provide means our ancestors would have dreamed of. Still we depict the earth round instead of elliptical or even not-symmetric since this kind of exactness is rarely helpful. Maps - or other models - cover different information like resources, population or like streets and houses. Even with space flight and the enormous telescopes today we deal with our universe in terms of simplified models.

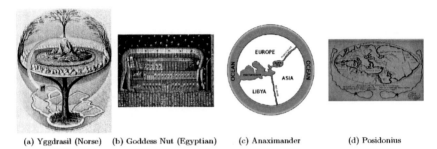

(a) Yggdrasil (Norse)   (b) Goddess Nut (Egyptian)   (c) Anaximander   (d) Posidonius

Figure 1.2: Cosmology - Models Of The World We Live In
(mythological and classical models)

In ancient times people also felt the need to explain their universe but without the modern tools the resulting models were bound to be less precise. Driven more by belief than scientific means the earth was depicted as a disc, and given that the physical laws were not known at that time, what other shape could the earth be? Observing objects on a flat table where everything stood in its place in contrast to a slope, the slope causes objects to slide downward. Centrifugal force was not an alternate idea then. Consequently the earth was supposed to be a flat shape in the middle of nowhere. Nowhere with no foundation feels ominous - down is the direction where objects tend to crash and nowhere is lots of down. A type of foundation must be devised to alleviate this threat. Different cultures developed diverse ideas how this foundation is supposed to look like. Germanic mystics thought that the universe is an incredibly large ash tree with discoid areas on its boughs. The central area is believed the home of humans and referred to by the name *midgard*, the topmost area *asgard* the home of the gods and *hel*, the underworld was at the roots (Figure 1.2a).

In ancient Egypt the goddess *nut* holds the heaven above the sky and protects the mortals from the chaos beyond (Figure 1.2b). Another idea was that the central area rooted on pillars of stone floating on an infinite ocean while other pillars held the heavens.

Many great spirits devised ideas how the universe is shaped, e.g. Anaximander (Figure 1.2c), Pythagoras, Plato, Posidonius (Figure 1.2d) and many others provided a more and more scientifically correct model - but this development took centuries.

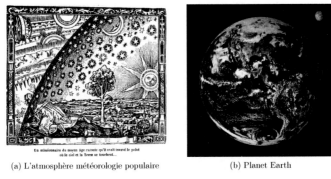

(a) L'atmosphère météorologie populaire　　　　　　　(b) Planet Earth

Figure 1.3: Cosmology - Models Of The World We Live In
(flat earth in contrast to the globe)

Thanks to today's technical gadgets nobody argues whether the earth is flat or not any more. Nonetheless flat earth served as an accepted model for many centuries.

1.2 Statistical Models

The ancient models mentioned above were derived by reasoning combined with known (or suspected) facts. Measurable facts can be subject to statistical modeling. The goal of such a statistical model is to gain inference from the measured data. This data are modelled as a structural relation between explanatory variables and a dependent variable, where the relation is blurred by noise.

Many alternative model forms have been proposed. Since noise and structure vary with the given data source there is no universal formula to fit a meaningful model. The distribution of the noise term as well as the scale of the measured data affect the model. Ideally the noise distribution can be identified so that the structural terms can be calculated correctly. A special subclass of statistical models are neural nets. These neural nets have been introduced to generate black-box models for forecasting events or recognizing patterns. Any black-box type of model rarely provides statistical inference and most neural nets are quite sensitive to the data they are trained with. This work will not cover black-box type methods but will give an overview of how statistic inference can be derived from data.

Other sciences depend on statistical model selection and use inference to find a better understanding of their own world. Financial analysts try to forecast share prices or future developments from companies, countries or other economic units. Jobs will be preserved or shed according to their analysis but usually there will be no immediate danger of life or death. Biologists, chemists or physicians might even use statistical inference to answer such critical questions. Biologists for example study animal or plant species and find answers about survival according to complex interactions. Chemists analyze how subatomic particles influence each other or discover new materials which improve life or can be deadly. Physicians use statistic models to find out how diseases can be diagnosed or how germs can be fought before life-threatening diseases break out.

As a matter of fact most modern sciences depends on knowledge gained from data - statistical inference. Psychiatrists, physicists, astronomers, sociologists, engineers and many more - all of them depend on derived statistical inference to explain how their world works.

Although statistical inference is a well known concept the way to gain it can be a tough one - even if knowledge is at hand before the actual modeling starts. The data to derive a model is only a small sample of reality and under most circumstances the experiment that yielded this sample cannot be recreated - or at least not without great efforts. So the data analyst must be very careful not to over-fit his models to fit the data perfectly since a perfect model cannot be used to filter out the structure of the problem. Over-fitting solves a numerical optimization problem but does not produce a good model. Noise is interpreted as structure and thus inference is based on spurious effects. The other extreme is not desirable either. When only the main structural effect is identified valuable structure remains hidden in the data. Under-fitting results in a model too general for explaining dependencies. If the distribution does not meet the assumptions, used for modeling the noise, any result is dubious.

1.3 Exploratory Model Analysis

Picking a good model is a delicate thing to do. Given a practical, often high dimensioned, data set it is a rare case that only one good model can be found. In a competition for the best model or set of candidate models either a selection rule must be specified or each model must be picked manually. Combining manual selection with interactive tools is a major part of exploratory model analysis.

Both types of investigation aim to find structure. Exploratory data analysis generates ideas while models are used for forecasting and tests. The data set is analyzed in the first step. After that models are fit. Sometimes one best model is identified. In the remaining cases a candidate set is chosen, often using interactive tools.

Exploratory data analysis precedes the model selection. First the data must be understood and outliers or other unusual features identified. Outliers are treated or even removed. The remaining observations are transformed into a final modeling form. In those cases where the appropriate transformation is not evident different kinds of transformation might be tried. Since a transformation skews the noise distribution any model selection must take care of any transformation applied. Models based on non-identically transformed variables must not be compared, especially not on condensed statistics like residual sums.

Correlation between explanatory variables may become evident during preliminary data analysis. In some cases a reduction of the eligible variables should be considered.

Dependent on the type of observation - categorical or numerical - the interactive data analysis steps save lots of time in choosing the right model type. This has been proven to be an invaluable help in the quest for valid model inference. A data analyst must explore the data before starting to explore models. Highly correlated variables lead to many related alternative models. Reducing variables minimizes model space - and exploring a huge model space consumes more time and effort than selecting a good set of predictor variables.

When the data examination has been finished models are fitted on the revised original data. Transformations offer a way to model non-linear effects as a simple linear model. The choice must be made by the analyst what type of model is favorable. Linear models are known for a long time and easier to understand and interpret than more complex model types. However, a transformation to coerce data to a form where a linear model can be fit is not generally advisable (see e.g Sen and Srivastava (1990)).

Each time when more then one dominating model can be identified the resulting models can be explored. This exploration might be done on a detailed level like residuals, or at a high level like model quality and complexity.

A short example will illustrate the idea. The data analysis part has been omitted on purpose to demonstrate on what characteristic models can be compared.

1.3.1 Example: Scottish Hill races

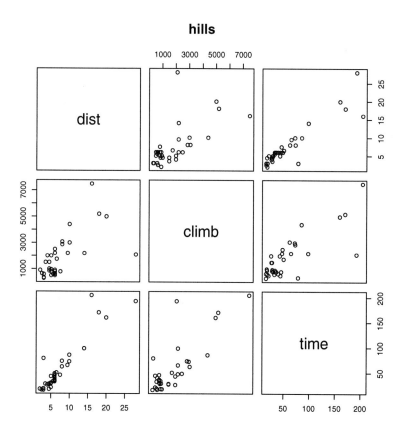

Figure 1.4: Scottish hill races data set, scatter plot matrix

Figure 1.5 shows the Scottish hills race data set as reported by Atkinson (1986b). This data set contains 35 records from Scottish hill races from 1984, featuring

climb the total height gained during the route, in feet,

dist the distance in miles (on the map) and

time the record time in minutes.

The data are highly correlated.

```
> cor(hills)
            dist      climb       time
dist   1.0000000 0.6523461 0.9195892
climb  0.6523461 1.0000000 0.8052392
time   0.9195892 0.8052392 1.0000000
```

The scatter plot (figure 1.4) does not apparently support a non-linear variable, so linear models are a valid choice:

Figure 1.5 shows the linear models where only one variable is picked for explanation. This facilitates the graphical presentation. No observations were removed though some observations require special treatment. Cook's distance, measuring the influence of an observation, suggests that some of the high leverage points like "Bens of Jura" should better be removed. Different transformations and the implications, even implications arising from removal of high influence points are discussed in Atkinson (1986b).

For convenience sake and to keep this example simple the hill races models are compared at different levels of detail:

overall goodness of fit The models are ranked on one global quality statistic. Typically AIC is used for linear models. Mean Squared Error (MSE), R^2 or the penalized version R^2_{adj} are other common options.

model parameters The structural parameters of the models are taken into account. More parameters produce better global goodness of fit performance but the principle of sparsity should be honored.

model residuals To find out how well the data sample is reproduced by the model residuals are compared. A parallel coordinate plot is a good option to compare residuals from many models but simpler plots might be helpful too.

The comparison strategies produce the following results:

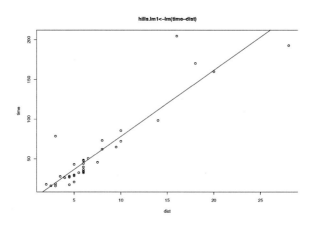

(a) Scottish hill race, scatter plot and linear model time vs. dist

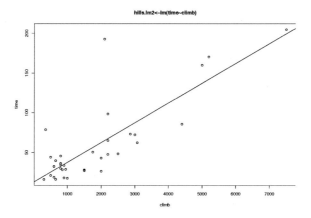

(b) Scottish hill race, scatter plot and linear model time vs. climb

Figure 1.5: hill race data, linear models

1.3.2 Global statistics

model	df	R^2	R^2_{adj}	AIC	deviance
time ~ climb	33	0.648	0.638	341.627	29933.83
time ~ dist	33	0.846	0.841	312.812	13141.61
time ~ dist + climb	32	0.919	0.914	292.222	6891.87
time ~ dist + climb + dist : climb	31	0.939	0.933	284.217	5178.42

Table 1.1: model comparison based on global statistics

Using this crude level of comparison the full model would be chosen by any of the quality statistics R^2 , R^2_{adj}, **AIC** or **deviance**. Of course the most accurate way to reproduce data with a model employs the use of every information available. At this level of detail there is no information if the parameters are valid or if the interaction term is required. Stepping deeper into the model details answers that question:

1.3.3 More specific statistics

```
Call:
lm(formula = time ~ dist)

Residuals:
    Min      1Q  Median      3Q     Max
-35.745  -9.037  -4.201   2.849  76.170

Coefficients:
            Estimate Std. Error t value Pr(>|t|)
(Intercept)  -4.8407     5.7562  -0.841    0.406
dist          8.3305     0.6196  13.446 6.08e-15 ***
---
Signif. codes:  0 '***' 0.001 '**' 0.01 '*' 0.05 '.' 0.1 ' ' 1

Residual standard error: 19.96 on 33 degrees of freedom
Multiple R-Squared: 0.8456,     Adjusted R-squared: 0.841
F-statistic: 180.8 on 1 and 33 DF,  p-value: 6.084e-15


Call:
lm(formula = time ~ dist + climb)

Residuals:
    Min      1Q  Median      3Q     Max
-16.215  -7.129  -1.186   2.371  65.121

Coefficients:
            Estimate Std. Error t value Pr(>|t|)
```

```
(Intercept) -8.992039   4.302734  -2.090   0.0447 *
dist          6.217956   0.601148  10.343 9.86e-12 ***
climb         0.011048   0.002051   5.387 6.45e-06 ***
---
Signif. codes:  0 '***' 0.001 '**' 0.01 '*' 0.05 '.' 0.1 ' ' 1

Residual standard error: 14.68 on 32 degrees of freedom
Multiple R-Squared: 0.9191,     Adjusted R-squared: 0.914
F-statistic: 181.7 on 2 and 32 DF,  p-value: < 2.2e-16

Call:
lm(formula = time ~ climb * dist)

Residuals:
    Min      1Q  Median      3Q     Max
-25.994  -4.968  -2.220   2.381  56.115

Coefficients:
              Estimate Std. Error t value Pr(>|t|)
(Intercept)  9.3954374  6.8790233   1.366  0.18183
climb       -0.0009710  0.0041648  -0.233  0.81718
dist         4.1489201  0.8352489   4.967 2.36e-05 ***
climb:dist   0.0009831  0.0003070   3.203  0.00314 **
---
Signif. codes:  0 '***' 0.001 '**' 0.01 '*' 0.05 '.' 0.1 ' ' 1

Residual standard error: 12.92 on 31 degrees of freedom
Multiple R-Squared: 0.9392,     Adjusted R-squared: 0.9333
F-statistic: 159.6 on 3 and 31 DF,  p-value: < 2.2e-16
```

Although the full model fits best and the parameter test statistic reports that the interaction term is highly significant the climb effect seems to vanish somewhere in this interaction. This full model is overfitted but the model using both parameters with no interaction receives good support from the test statistic for all effects and good quality as well. Additionally any combined effect like climb and distance is more difficult to explain - especially since the climb effect does not receive good support.

Using different views on model residuals identifies the high leverage points and provides support for a normally distributed error.

1.4 Model Architecture

The introductory example showed how a linear model could be used to describe the data, more precisely the relation between data. Since no data cleaning has been performed there is still room for improvement in terms of better model quality.

Considering more complex data the concept of *parsimony* is important. Since the quality can be improved by adding more parameters, when is the best point to stop? If models grows too complex the given structure cannot be clearly interpreted. The principle of *parsimony*, as proposed by Box and Jenkins, suggests that the model with "... the smallest possible number of parameters for adequate representation of the data" should be used.

All of the above holds true for the case of linear models and more complex types of models. Linear models are the easiest to interpret and - given that the preconditions hold true - should be used. By examining the data before fitting models the analyst carefully selects a valuable subset of the available data. If data become sparse or it will be infeasible for various reasons to get more data the model selection will be more challenging.

1.4.1 Model Selection

When a very complex model is used to describe the data, this model is presumably over-fit, which means that the model is too specific to be used for another data sample save the used one. If the model has too few parameters the model is not of much use either since structural information is missing. The balance between complexity and interpretability (in most cases this is a synonym for parsimony) needs to be kept. The term *bias-variance tradeoff* is related to the problem at hand. In James and Hastie (1997) this prediction error was shown to be representable by the variance and the bias term, even for general random variables. Equation 1.1

$$\underbrace{EL_S(Y,\hat{Y})}_{\substack{\text{Prediction}\\\text{Error}}} = \underbrace{EL_S(Y,SY)}_{\substack{Var(Y)\\\text{irreducible}\\\text{Error}}} + \underbrace{E\left[L_S(Y,S\hat{Y}) - L_S(Y,SY)\right]}_{bias^2(\hat{Y},SY)} + \underbrace{E\left[L_S(Y,\hat{Y}) - L_S(Y,S\hat{Y})\right]}_{Var(\hat{Y})} \quad (1.1)$$

$$\underbrace{\phantom{E\left[L_S(Y,S\hat{Y}) - L_S(Y,SY)\right] + E\left[L_S(Y,\hat{Y}) - L_S(Y,S\hat{Y})\right]}}_{\text{reducible Error}}$$

implies that prediction accuracy can be increased by introducing a higher model bias - the *bias variance tradeoff*. In practice that means that a too specific model using too many parameters will have a high accuracy though this accuracy is skewed by the bias with respect to the used sample.

Usually only one sample is at hand. This fact leads to practical problems, as the same sample is used to fit any model. Either model space is searched exhaustively, or some heuristic algorithm is applied. Stepwise procedures are used to select variables in practice, though these are subject to a well-known set of biases.

1.4.1.1 Model Selection Bias

Any incremental or stepwise procedure suffers from bias at the point of selecting the follow-up model. On the one hand the same data sample is reused for calculating the subsequent model. This leads to overly optimistic measures according to Burnham and Anderson (2010). On the other hand many alternative variable combinations are left out of consideration. The last model is dropped in favor of the better alternative and lost in that process; many possible alternatives are not even considered for selection. Breaking down the stepwise procedure different kinds of bias can be distinguished. In his paper Miller (1984) shows three different types of model selection bias, namely a bias due to

- the omission of variables,
- the competition for selection and

- the stopping rule.

Miller shows these bias terms in the context of linear regression. Formally the data sample is partitioned in two subsets $\mathcal{D} = (\mathcal{D}_A, \mathcal{D}_B)$ as well as the regression coefficients $\beta = (\beta_A, \beta_B)$. The data is partitioned into distinct subsets of the variables while the sample is not altered. The expected values for b_A from least squares, with X_A, X_B as the partitioned data matrix will be

$$E(b_A) = \beta_A + (X_A^T X_A)^{-1} X_A^T X_B \beta_B \tag{1.2}$$

The right hand term $(X_A^T X_A)^{-1} X_A^T X_B \beta_B$ in formula 1.2 can be interpreted as the bias of *omitting* the variables from subset B.

Assuming that the subset A has been chosen for an optimal model fitting the expected predictors in $E(b_A)$ will in general be different from 1.2. This difference is called the *selection bias* and, if these subsets yield different results, the kind of bias is called *competition* bias. The *stopping rule bias* results from the fact that some iterative algorithm that reuses the same data will stop when the stopping rule applies. Neither the algorithm nor the stopping condition can guarantee the best possible variable selection.

1.4.1.2 Model Selection Uncertainty

Another problem with model selection is the *model selection uncertainty*. This *uncertainty* relates to the sample covered by the available data set in contrast to other samples. Any provided data set is just a finite, usually small, sample of a whole population - reality. The real distribution can only be captured roughly by any finite sample. Some aspects of a whole population cannot be captured but some extraordinary observation might be found in that random sample. Since the model variables are fit to the sample a common challenge is to pick the best possible variables and identify valid parameter estimates. Data set related *uncertainty* affects the fitted model. In rare cases some dominant, high-leverage, observation can skew the variable selection process in a very undesirable way. Assume that the sub-sample $\mathcal{D}_S = \mathcal{D}_{s(1)..s(m)}$ from the whole data set $\mathcal{D} = \mathcal{D}_{s(1)..s(n)}$, with $m \ll n$. The leverage induced from $\mathcal{D}_S$ will provide a model that fits the variable v_κ if $\mathcal{D}_S$ is present in the data but if the sub-sample $\mathcal{D} \setminus \mathcal{D}_S$ is used no model with v_κ will be found. In that way the concept of model selection bias is related to the selection uncertainty. In contrast the question here is if the model from one finite sample fits another sample from the same population equally well.

In other words the candidate set or the single best model that has been picked from a finite sample will provide parameter estimates with respect to that very special sample used; different samples from the same population will provide other data that are fitted to a slightly different model. Parameter estimates differ, although the difference (hopefully) is bounded by a small interval. Nevertheless the sample could be picked so unfortunately that fitted models differ to such an extend that forecasts are invalid. From the point of variable identification the fact that different variables are selected for the best model is disturbing. Different samples from the same population yield different parameter estimates and in some cases even the selected variables can vary.

Another possible cause for selecting a sub-optimal subset of variables is the algorithm that selects the predictor variables. This is referred to as model variable selection uncertainty. Miller (1984) points out that for the DETROIT data set while using forward selection the same variable is selected first as will be dropped out by backward selection. Another example used in this paper shows an artificial example from Berk where both forward and backward selection result in the same variable selection but both miss the best possible model - better by a remarkable amount. Some techniques based on the bootstrap methods have been explored and will be discussed later on that reduce the risk of variable selection uncertainty dramatically.

1.5 Summary and Outlook

Model variable selection is not a simple task for any unsupervised algorithm. Unsupervised means a computer is automatically processing a part of model space and the result is not validated manually. Nonetheless automatic stepwise selection is used regularly and best models are used for inference. Lots of variable combinations are never used, others are disregarded in search of only one best model. This approach wastes lots of information contained in the data set. Measuring information as a general framework for model comparison is introduced in the next chapter.

To counter too great information loss a valid approach is to keep all good models for inference. The mixed effects are harder to interpret than any single model, but on the positive side more information is preserved. Model space is infinite, so finding a good set of models is generally costly. The reduction in terms of bias and accuracy is often worth these efforts. Recent methods employing model ensembles are discussed in the following chapter.

Because unsupervised algorithms optimize with respect to one defined criterion merely suboptimal models are fit. Manual supervision is costly. A short excursion into interactive methods is presented as a means for manual supervision. Model ensembles are not data sets in the usual sense, but derived data. Nevertheless the amount of derived data requires interactive methods for manual supervision.

Manually managing model sets involves lots of effort. Possibly many statisticians shy away from the effort of handling lots of models. As long as manual management, which is laborious, is the only possibility of handling model sets this attitude is understandable. So a logical step is to provide means that to make task less cumbersome. In order to achieve that goal a computer program is the most powerful aid any statistician can wish for. Models have been fit using computer programs for many years now. Yet managing lots of models is still a neglected feature in modern statistical software. What features are required and useful for model management?

To answer this question a software has been created to manage lots of models. But model management left alone is only a minor benefit for an analyst. So examples and tools are presented in the following chapters to demonstrate what a software for model ensemble support should be able to do. The prototype has been build upon the model facilities of R (R Development Core Team, 2006), leveraging a large user community with model building knowledge and many other helpful features.

Finally the results and experiences are summarized again, highlighting some of the innovations and giving directions for further research.

Chapter 2

Model Comparison

2.1 Introduction

To find a good model or set of models these models must be selected in some form of competition. Supposed the data sample contains all characteristics that could be observed, the full model is a natural competition champion. Whether such a model can be constructed from a finite sample is dubious. The existence of such a *true model* is theoretically possible but unattainable in practice. A finite sample exhibits the amount of reality that could be captured on a sensible basis, so this sample must stand in for full reality. Any model constructed is fitted on that special sample and is only able to reproduce features found in that sample. Self-evidently no structure can be modeled from data where no observation that supports that structure is found in the data set. (see §1.4.1.2 on p. 11)

Abstraction or simplification from the sample results in further loss of information and a measurement of this loss is the *Kullback-Leibler Information* §2.2. Maximum likelihood is presented as a general framework where ordinary least squares estimation is a special case for finding best approximating distributions or models.

In practice models are compared and ranked by global criteria. Linear models support a wide range of global criteria - one takes only regard of model complexity. Another few are solely based on model quality and many more take both into account. In the introductory example (table 1.1, p. 8) the linear models have been compared on df[1] that takes only model complexity into account. R^2 only evaluates the model residuals and thus only supports the model quality. R^2_{adj}, AIC and *deviance* use different combinations of the above. This did not suffice and more detailed information from the coefficients was needed for the model selection. One problem became obvious: selection of a model solely on the base of one global criterion is misleading. Partially this problem arises from the paradoxical situation that model fit and thus quality will always be improved on complexer models - yet model complexity is not a desired quality in a model. Following both objectives, to achieve a good fitting model with low complexity is a primary goal. To achieve this goal lots of so called *information criteria* are proposed in the literature. §2.2.2 covers an extensive list of these criteria as well as some thoughts about the usefulness of these criteria for practical model selection.

---

[1]Never use df as a selection criterion alone. The null model wins that competition and no inference may gained that way.

2.2 A distance between models

Simplification from reality results in the loss of information. Any distribution in reality, a finite sample or even a model can be interpreted as a continuous function. To compare such functions and find a measure for this information lost the Kullback-Leibler Information or *discrepancy* is a universal tool - discrepancy because it is not a distance function in the original sense as this measurement is not symmetric.

Definition 1 *Let f and g be two continuous functions, for example probability distributions or models.*
The **Kullback-Leibler Information** *between two models is defined as*

$$I(f,g) = \int f(x) ln \left(\frac{f(x)}{g(x|\theta)} \right) dx$$

where $g(x|\theta)$ is a distribution dependent on a parameter vector θ

In the discrete case the distance can be written as

$$I(f,g) = \sum_{i=1}^{k} \left(\frac{p_i}{\pi_i} \right)$$

where $p_i = P_f(X = i)$ represents the probability under f and $\pi_i = P_g(X = i)$ the probabilities under g.

The Kullback-Leibler Information can be interpreted as the information lost when g is used to approximate f.
Since these probability distributions are usually not available straightforwardly the concept of relative distance is used in practice:

$$
\begin{aligned}
I(f,g) &= \int f(x) ln\left(f(x)\right) - \int f(x) ln\left(g(x|\theta)\right) \\
&= E\left[ln\left(f(x)\right)\right] - E\left[ln\left(g(x|\theta)\right)\right] \\
&\vdots \quad \text{see Bozdogan (1987) for detailed derivation} \\
&= C - E\left[ln\left(g(x|\theta)\right)\right] \\
I(f,g) - C &= -E\left[ln\left(g(x|\theta)\right)\right]
\end{aligned}
$$

2.2.1 Finding the best model

The information loss measure provides means to compare models based on the assumption that the true probability distribution is known in advance. Before models are compared the first step is to fit models to the finite sample of a given data set. In practice the real distribution of the modeled data is usually unknown but after some transformations and data exploration the assumption of normality will often be made. Under these conditions models can be fit by the method of ordinary least squares. If the assumption of normality holds true, the ordinary least squares results equal the results gained from the maximum likelihood estimator.

2.2.1.1 Ordinary Least Squares

The ordinary least squares (OLS) estimate for a regression model

$$y = X\beta + \varepsilon$$

where y is the variable to be explained, X a matrix representation of the data set and ε the residuals, can be found by minimizing

$$(y - X\beta)^t(y - X\beta)$$

Assuming that the matrix X^tX is non singular a unique solution for β is given by $(X^tX)^{-1}X^ty$. Linear dependent variables are left out after the data analysis step. A solution may be numerically calculated using the Moore-Penrose Inverse even when the model matrix is singular.
Given the Gauss-Markov conditions

$$
\begin{aligned}
E(\varepsilon) &= 0 \\
E(\varepsilon\varepsilon^t) &= \sigma^2\mathcal{I} \text{ (where } \sigma^2 < \infty \text{)}
\end{aligned}
$$

hold true the ordinary last squares estimate for β provides the best linear unbiased estimator (BLUE). There are lots of books about OLS estimation. For linear regression Sen and Srivastava (1990) provides a nice introduction.

2.2.1.2 Maximum Likelihood

The maximum likelihood method was discovered by Carl Friedrich Gauss in 1821 and refined in 1922 by Sir Ronald Aylmer Fisher (see Aldrich (1997) and bibliography). Given a parameter vector θ the likelihood function can be written as

$$\mathcal{L}(\theta|x) = \prod_{i=0}^{n} p(x_i|\theta)$$

assuming that the observations are independent and from the same distribution family. This equation can be read as the likelihood that an observation x is made under the parameters θ where p is a probability density function. Many useful distribution families have the nice feature that they are unimodal. Additionally many are even convex. This guarantees that a unique maximum for this likelihood function exists and the solution can at least be numerically calculated. If a unique solution for the likelihood equation can be found this solution is used for parameter estimation. In practice the log-likelihood will be used in most cases because it is more convenient to handle:

$$\ell(\theta|x) = \sum_{i=0}^{n} \ln\left(p(x_i|\theta)\right)$$

Some properties of this likelihood functions are functional invariance, asymptotic unbiasedness, asymptotic efficiency and asymptotic normality. More details can be found in Davison (2003) or a general short and nice introduction to likelihood in the appendix of McCullagh and Nelder (1989).
Under mild conditions i.e. that a maximum can be found, the term maximum likelihood is used

for the obtained optimal solution.

Another interesting term related to the likelihood equation is the *information* from the fisher information matrix. The *observed information* is the negative Jacobean of the log-likelihood

$$J(\theta) = -\frac{\partial^2 \ell(\theta)}{\partial \theta^2}$$

The *expected* or *Fisher information* is defined as

$$I(\theta) = -E\left\{\frac{\partial^2 \ell(\theta)}{\partial \theta^2}\right\}$$

This term will be encountered in the next section about information criteria.

2.2.2 Information Criteria

An information criterion is a single statistic that summarizes the overall quality of a model. Since the quality in terms of "what model fits the data best" can be expressed by the measurement of the residuals alone even σ^2 might be used for such a purpose. The drawback of this criterion is that the full model is chosen as the "best" one. Selecting models only with respect to their quality and disregarding the complexity will result in too specific (over-fitted) models, optimized on a special data set. No generalization can be made. The chosen model is too complex and not interpretable.

The key to solve this dilemma is to chose a model on the basis of good quality but also pay attention to the model complexity. The principle of parsimony is respected this way. In order to use a single value - the information criterion - for model selection the complexity of the model is used to penalize the quality. How much penalty should be used is still a matter of discussion - the theoretical background how to chose a penalization can be derived from the maximum likelihood principle (see e.g. Burnham and Anderson (2010)) but other theories have been explored as well. Consequently an enormous number of criteria have been proposed.

AIC, TIC are directly derived from maximum likelihood theory and many subtypes have been explored that take special account of the size of the data set.

A completely different approach is taken in statistical learning theory like VC-Dimension, SRM etc.

From the computer science and encoding theoretic background MDL, MML and similar criteria are derived.

An interesting fact is that some of these criteria have the same asymptotic behavior, like *TIC* and *AIC*. It is nice to know that on a basis of very large data sets the results are likely identical. Unfortunately in practice data sets can be annoyingly small and more data is hard or even impossible to come by. No universally superior criterion has been identified. So a feasible criterion must be selected case by case. The analyst must select the best model or set of candidate models dependent on the data set and may chose any criterion that matches the given circumstances and not vice versa.

Table (2.1) on the following pages is not complete and does not present every single information criterion ever published but it pools a wide range of ideas. Due to the different sources and theoretic backgrounds the notation is not normalized and many of the displayed criteria can be used in different types of applications - e.g. *AIC*.

These criteria were selected over various applications that relate to statistical modeling. E.g.

time series analysis, regression, learning theory and many more. Nevertheless it is quite remarkable that so many criteria has been published.

Counting 71 different criteria from table 2.1 the usefulness or need for new criteria is doubtful. Many "new" criteria are derived from existing and theoretically well-founded criteria like *AIC* with just a special modification to the penalization term. This modified criterion choses a superior best model with respect to one special data set.

While it is useful to observe the influence of data size versus asymptotic behavior the combinations for the penalization term are virtually unlimited. A simple note why the result was not the best model due to some known criterion with a sound justification will yield more insight than the creation of another specialized criterion could possibly provide.

Modern computers can cope with an arbitrary amount of criteria for any model. But there is no reason to compute as many criteria as possible. Not the majority vote over a set of many criteria yields good model selection. One or, in rare cases, few criteria, separate good models from the rest. The only use of information criteria is to chose an eligible set of models. If criteria disagree about the selected models, a manual selection must be made.

Criterion	Reference	Formula
A_p	Allen (1974)	$\sum_{i=1}^{n}\left(\frac{\varepsilon_i}{1-h_{ii}}\right)^2$ (see PRESS)
AIC	Akaike (1974)	$-2\ln(\mathcal{ML}) + 2p$
AIC_C	Hurvitch and Tsai (1989)	$-2\ln(\mathcal{ML}) + 2p + \frac{2p(p+1)}{n-p-1}$
AIC_U	McQuarrie and Tsai (1998)	$-2\ln(\mathcal{ML}^\star) + 2p + \frac{2p(p+1)}{n-p-1}$
$AICW$	Wilks (2005)	$-2\ln(\mathcal{ML}) + 4p$
$BCIC$	Bozdogan (2000)	$-2\ln(\mathcal{ML}) + 2nb$
BIC	Schwarz (1978)	$-2\ln(\mathcal{ML}) + 2p\ln(n)$ (see SBC)
C_p	Mallows (1973)	$\frac{RSS}{s^2_{full}} - n + 2p^\alpha$
$CAIC$	Bozdogan (1987)	$-2\ln(\mathcal{ML}) + p[\ln(n)+2] + \frac{p}{2}n$
$CAICF$	Bozdogan (1987)	$-2\ln(\mathcal{ML}) + p[\ln(n)+2] + \ln|\hat{J}|$
CAT	Parzen (1974)	$\frac{1}{n}\sum_{j=1}^{p}\frac{1}{\hat\sigma_j^2} - \frac{1}{\hat\sigma_p^2}$
$CAT_\alpha(k)$	Bhansali (1986)	$1 - \frac{\hat\sigma^2}{\sigma^2(k)} + \alpha\left(\frac{k}{n}\right)$
$cSIC$	Sugiyama and Ogawa (2001)	$max\left(0, \|\hat{f_\theta} - f_u\| - trace(X_0 Q X_0^t)\right) + trace(X_0 Q X_0^t)$
$CIC(\kappa)$	Tibshirani and Knight (1999)	$\overline{err}(\kappa) + \frac{2\hat\sigma^2}{n\hat\sigma_y^2}\sum_{i=1}^{n}Cov^0(Y_i \eta_i \star (x_i, M_\star^\star)) + \frac{2}{n}\hat\sigma^2$
DIC	Spiegelhalter et al. (2002)	$-2\ln((\mathcal{ML}(y|\theta)) + C$
EIC	Ishiguro et al. (1997)	$-2\ln(\mathcal{ML}) + 2E_{X^\star}\cdot\left\{\ln f\left(X^\star|\hat\theta(X^\star)\right) - \ln f\left(X|\hat\theta(X^\star)\right)\right\}$
FIC	Claeskens and Hjort (2003)	$\omega^t(I - K^{\frac{1}{2}}H_S K^{-\frac{1}{2}})DD^t(I - K^{\frac{1}{2}}H_S K^{-\frac{1}{2}})^t\omega + 2\omega^t K^{\frac{1}{2}}H_S K^{-\frac{1}{2}}\omega$
$FIC(\kappa)$	Wei (1992)	$n\hat\sigma_p^2 + \sigma^2\ln\left|\sum_{i=1}^{n}x_i(\kappa)x_i^t(\kappa)\right|$
FPE	Akaike (1973)	$\frac{RSS(n+p)}{(n-p)}$
$FPE(\lambda)$	Akaike (1970)	$ASE(\lambda)\left(\frac{1+\frac{Q(\lambda)}{n}}{1-\frac{Q(\lambda)}{n}}\right)^2$
FPE_u	McQuarrie and Tsai (1998)	$\frac{RSS n(n+p)}{(n-p)^2}$

---

[2]from Moody (1994)

$FPE\alpha$[3]	Bhansali and Downham (1977)	$\frac{RSS(n+3p)}{(n-p)}$
$FPEC$	de Luna (1998)	$\hat{\sigma}_p^2\left(1 + 2\frac{p}{n}\right)$
$FPECR$	Johansen (1998)	$p = argmax_k \hat{\sigma}_k^2\left(1 + \frac{k}{n}\right) + \frac{\hat{\sigma}_i^2}{n}\chi^2_{k-m}(1-\alpha) < \hat{\sigma}_m^2\left(1 + \frac{m}{n}\right)$
$FPER$	?	$FPECR(\gamma; Z^N) = \frac{1 + \frac{d_v(\gamma)}{N}}{1 - \frac{d_v(\gamma)}{N} + \frac{d_i(\gamma)}{N}} \cdot V_N(\hat{\theta}_{REG,\gamma}; Z^N)$
$GAIC$ (TIC)	Takeuchi (1976)	$\frac{N + \hat{m}_2}{N - 2\hat{m}_1 + \hat{m}_1} S_n(\hat{\omega})$
$GCV(\lambda)$	Craven and Wahba (1979)	$-2\ln(\mathcal{ML}) + 2trace(\hat{J}(\theta)^{-1}\hat{I}(\theta))$ see TIC $ASE[\lambda]\frac{1}{\left(1 - \frac{Q(\lambda)}{n}\right)^2}$ [4]
GIC	Nishii (1984)	$-2\sum_{i=1}^n \ln f_\theta(z_i) + \frac{2}{n}\sum_{i=1}^n trace\left(t^{(1)}(z_i, G)\frac{\partial \ln f_\theta(z_i)}{\partial \theta}\right)$
GIC_2	Kitagawa and Konishi (1999)	$-2\ln(\mathcal{ML}) + 2\left\{b_1(\hat{G}_n) + \frac{1}{n}\left(b_2(\hat{G}_n) - \Delta b_1(\hat{G}_n)\right)\right\}$
GM	Geweke and Meese (1981)	$\frac{RSS}{s_{full}} - p\ln(n)^\alpha$
$GPE(\lambda)$	Moody (1991)	$\varepsilon_{train} + \frac{2}{n}trace\hat{V}\hat{G}(\lambda)$
$GSIC$	Tsuda et al. (2002)	$(\theta - \theta^u)^T P(\theta - \theta^u) + 2\sigma^2 trace(PW^0) - \sigma^2 trace(PV)$
HAR	Haring (1975)	$min\left(-(n - m)\ln\left(\frac{(n-m)\mathcal{E}(m+1)}{(n-m+1)\mathcal{E}(m)}\right)\right)$
HQ	Hannan and Quinn (1979)	$-2\ln(\mathcal{ML}) + 2p\ln(\ln(n))$
HQ_C	McQuarrie and Tsai (1998)	$n\log\left(\frac{RSS}{n}\right) + \frac{2pm\log\log n}{n-p-2}$
$ICL_{m,K}$	Biernacki et al. (2000)	$-2\ln(\mathcal{ML}, \mathcal{K}) - \frac{\nu_{m,K}}{2}\ln n$
$ICOMP$	Bock (1988)	$-2\ln(\mathcal{ML}) + p\ln\left(\frac{trace\hat{\Sigma}_p}{p}\right) - \ln|\hat{R}_{pl}|$ $+ n\ln\left(\frac{trace\hat{R}}{n}\right) - log|\hat{\Sigma}_{pl}|$
$IFIM$	Bozdogan (1990)	$I_n(\theta)_{i,j} = cov\left[\frac{\partial \ln f(X|\theta)}{\partial_i}, \frac{\partial \ln f(X|\theta)}{\partial_j}\right]$
JEW	Jenkins and Watts (1968)	$min\left(\mathcal{E}\frac{n-m}{(n-2)*(m-1)}\right)$
K_C	Kashyap (1982)	$\ln f(x|\theta)\frac{1}{2}\ln h(\theta) + \frac{p}{2}\ln(n) + \frac{1}{2}\ln|B(\theta)|$
KIC	Cavanaugh (1999)	$-2\ln(\mathcal{ML}) + 3(p+1)$
KIC_c	Cavanaugh (2004)	$-2\ln(\mathcal{ML}) + n\ln\frac{n}{n-p} + \frac{n((n-p)(2p+3)-2)}{(n-p-2)(n-p)}$
$LEIC$	Billah et al. (2003)	$-2\ln(\mathcal{ML}) + kp$

[3] also found as $FPE4$
[4] from Moody (1994)

$MAIC$	Fujikoshi and Satoh (1997)	$-2\ln(MC) + \frac{2n(p+1)}{n-p-2} + 2p\left(\frac{(n-p)\hat{\sigma}_p^2}{(n-P)\hat{\sigma}_P^2} - 1\right)$ $-2\left(\frac{(n-p)\hat{\sigma}_p^2}{(n-P)\hat{\sigma}_P^2} - 1\right)^2$
MEM	Jaynes (1957)	$MEM = argmax_{p \in C} - \sum_{x,y} \bar{p}(x)p(y|x)\ln p(y|x)$
MDL	Rissanen (1978)	$MDL(x^n|p) = -\ln P(x^p|\hat{\theta}) + \frac{p}{2}\ln n$
$MKIC$	Cavanaugh (2004)	$-2\ln(MC) - \frac{2n(p+1)}{n-p-2}$
MML	Wallace and Dowe (1999)	$-\ln\left(\frac{h(\theta)p(x|\theta)}{\sqrt{F(\theta)}}\right) + \frac{1}{2}\ln(\frac{c}{12})$
NIC	Murata et al. (1994)	$D(q^*, p_i(\hat{\theta}_i))) + \frac{1}{t}trace\left(G_i^*\hat{\theta}_i Q_i^*(\hat{\theta}_i)^{-1}\right)$
$NLEIC$	Billah et al. (2003)	$-2\ln(MC) - k_{pp}$
PHI	Pukkila and Krishnaiah (1988)	$-2\ln(MC) - 2p\ln(\ln(n))$
$PMDL$	Rissanen and Ristad (1994)	$PMDL(x^n|k,d) = -\sum_{t=0}^{n-1}\ln P(x_{t+1}|x^t, \hat{\theta}(x^t))$
$PRESS$	Allen (1974)	$\sum_{i=1}^{n}\left(\frac{\varepsilon_i}{1-h_{ii}}\right)^2$ (see A_p)
$PSE(\lambda)$	Barron (1984)	$ASE(\lambda) + 2\hat{\sigma}^2\frac{Q(\lambda)}{n}$
$QAIC$	Lebreton et al. (1992)	$-2\frac{\ln(MC)}{\hat{c}} + 2p$
$QAIC_C$	Lebreton et al. (1992)	$-2\frac{\ln(MC)}{\hat{c}} + 2p + \frac{2p(p+1)}{n-p-1}$
QPK_k	Pukkila and Krishnaiah (1988)	$\ln|\hat{\Sigma}| + \frac{kd^2 + kd(\ln(T)-1)}{T-0.5d(k+1)}$
R_p	Breiman and Freedman (1983)	$\frac{RSS(n-1)}{(n-p)^2}$
RIC	Foster and George (1994)	$-2\ln(MC) + 2n\ln(p)$
$RMSPE$	Feng and Liu (2003)	$\sqrt{\frac{\sum_{i=1}^{n}\left(\frac{\varepsilon_i}{y_i}\right)^2}{n}}$
SBC	Schwarz (1978)	$-2\ln(MC) + 2p\ln(n)$ (see BIC)
SIC	Sugiyama and Ogawa (2001).	$(\hat{\theta} - \theta^u)^T P(\hat{\theta} - \theta^u) + 2\sigma^2 trace(PW) - \sigma^2 trace(PV)$
$SH(\kappa)$	Shibata (1980)	$argmin_{M_n}\hat{\sigma}_\kappa^2 \frac{(n+2k)(n-k)}{n}$
S_p	Stine (2004)	$\frac{RSS}{s_{full}^2} + \hat{\sigma}^2 + \sum_{i=1}^{p}j\delta(j)$
SRM	Vapnik (1998)	$min(R_n(g) + B(h,n))$
TIC	Takeuchi (1976)	$-2\ln(MC) + 2trace(\hat{J}(\theta)^{-1}I(\theta))$

WIC	Ishiguro and Sakamoto (1991)	$-2\ln f(y\|\hat\theta_k) + \left\{ \dfrac{1}{B}\sum_{i=1}^{B} -2ln\,\dfrac{\ln f(y\|\hat\theta^*_k(i))}{\ln f(y(i)\|\hat\theta^*_k(i))} \right\}$
VC − dimension	Vapnik and Chervonenkis (1971)	$\hat R(f_n) = R_{emp}(f_n)\left(1 - c\sqrt{\dfrac{d(1+\ln\frac{n}{d})-\ln\delta}{n}}\right)^{-1}_+$
WAIC	Irizarry (2001)	$-2\ln(\mathcal{ML}) + 2trace(\hat i(\theta)\hat J(\theta)^{-1})$
WBIC	Irizarry (2001)	$-2\ln(\mathcal{ML}) + \ln\det(-\hat J)$
WCAICF	Irizarry (2001)	$-2\ln(\mathcal{ML}) + 2trace(\hat i(\theta)\hat J(\theta)^{-1}) + \ln\det(-\hat J)$

Table 2.1: A list of information criteria.

Other ways to chose the best model - or respectively employing stepwise selection - are

- cross-validation
- double cross-validation
- posterior odds (Zellner, 1996)

This list of criteria (table 2.1) is not complete and is not meant to be. Lahiri (2002) contains a nice summary of many criteria and their application (pages 7 -59). Additional criteria found there involve for example *PLS*, *MIC*, *AICR*, RC_p, *AICcR**, *AICcR*. McQuarrie and Tsai (1998) compiled an extensive list of comparison and simulation results.

Curiously the number of published criteria is raising. Most new publications deal with special data set related complexity issues which means that the penalty term is varied accordingly.

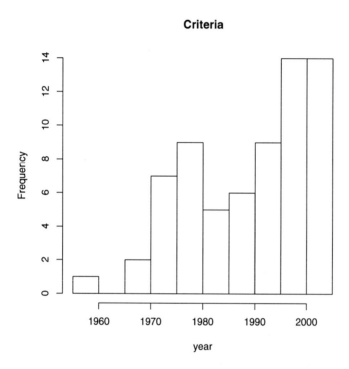

Figure 2.1: criteria published per year

Chapter **3**

Model Selection Strategies

The competition winner in the quest for a good model can be found using one of the many information criteria presented in table 2.1 (p. 21). But even the top contenders found are not automatically guaranteed to serve well as a model for reality. Partially this issue arises from the model selection problems described in §1.4.1. The selection procedure is subject to bias from different sources. Model selection bias is introduced by using only a subset of the variables. The different competition situations encountered during the selection procedure are biased in the way that the defined stopping rule will arbitrarily quit and thus stop the creation of untested valid models. The data sample collected from reality might be poorly chosen or badly conditioned, introducing a further source of spurious effects that results in selection uncertainty. Some difficulties selecting models are hardly avoidable - still some practical solutions have been found to counter poor model selection. One idea is to partition the data set for evaluation purposes. Another consideration is not to rely on a single best model but to collect all good models into a candidate set and use the combined information.

3.1 Model Validation

The data available for analytical purposes is only a finite sample of reality. In practice there is seldom a chance to collect another sample, to validate the model selection result. Any data sample contains errors from different sources, like imprecise measurement for instance. If the sample size is large enough measurement should not pose an obstacle but given many practical problems it is not feasible to obtain large samples. In practice the unique data set available contains structural data, noise but also spurious effects. These spurious effects should not be included in the candidate models because inclusion means there is a non-structural factor included, special to the data set which is equivalent to over-fitting the data. High leverage points and influence can pose another problem. To cope with data related problems the basic idea of model validation is to partition the data set into a training set and a test set. The model is computed from the training set and this result is validated against the test data.

3.1.1 Cross-Validation

A practical method of validation developed by Stone (1977) is the *loo* method for *leave one out* cross-validation. One observation is removed from the data set and compared against a reference forecast from the model. The model is fitted on the reduced data set. Cross-validation implies that the model is calculated n times and each time a different observation is left out. Model selection is based on average validation error.

Given enough data the validation partition might be chosen larger and the model can be recalculated using another selection of training observations. Instead of *leave one out* this method is referred to as *leave many out* or *PSR* for predictive sample reuse and described by Geisser (1975). The data set $\mathcal{D}$ containing n observations is partitioned into a training $\mathcal{D}_{TRAIN}$ and validation part $\mathcal{D}_{VAL}$. Any type of cross validation will use $\frac{n}{|\mathcal{D}_{VAL}|}$ validation steps cycling through the training data $\mathcal{D}_{TRAIN} = \mathcal{D} \setminus \mathcal{D}_{VAL}$. In practice these subsets are partitioned once into fixed equal parts instead of calculating every possible combination.

3.1.2 Resampling Techniques

Partitioning the data set for validation purposes is one way to take advantage of sub-samples from the original data. Resampling - as this technique is called - can be useful for bias evaluation or to derive estimates of other properties regarding the sampling distribution. Resampling means that the original data set is modified - usually a subset of the original data is drawn. The most simple type is leaving one observation out[1]: this is abbreviated to *loo* or alternatively called the *Jackknife* (Quenouille, 1956). Leaving out many observations is viable, too.

Another possibility is to use samples of the same size as the original data but to replace the observations. This is achieved by drawing a random index sample and fitting the indexed observations. The original data would be referred to as $\mathcal{D}_1, \mathcal{D}_2, \ldots \mathcal{D}_n$. Let $B : \{1 \ldots n\} \mapsto \{1 \ldots n\}$ be a random index function. Then $\mathcal{D}_{B(1)}, \mathcal{D}_{B(2)}, \ldots \mathcal{D}_{B(n)}$ is a sample with replacement. The Bootstrap(Efron, 1987) uses a number of sub-samples and has a wide range of applications.

The *Jackknife* was introduced by Quenouille (1956) to estimate the bias of the sampling distribution. One observation is left out from the assessed sub-sample like in the *loo* cross validation method and the n sub-samples are averaged for bias or variance estimation. For model averaging concerns a technique named *jackknife model averaging* has lately been developed by Hansen and Racine (2007). This frequentist method is asymptotically optimal and does not preclude a heteroscedastic setting. Even the finite-sample performance outmatched many other methods.

The bootstrap choses observations randomly while allowing duplicate observations. The drawing process can either be parametric - so the samples are drawn from a fixed distribution - or non-parametric. Efron and Tibshirani (1994) gives an overview of different applications of the bootstrap technique in chapter 7 on model assessment and chapter 8 about the relation to the maximum likelihood principle. Kohavi (1995) provides a comparison of different cross validation techniques.

---

[1] resampling is a more general framework than cross validation, so loo is mentioned here, too

3.2 Single Model Selection

Single model selection has been successfully applied for a long time and different kinds of selection strategies and stopping rules were used to select one model for inference. Forward variable selection or backward variable elimination or mixtures were used to compute the best model step by step. See for instance Venables and Ripley (1998) or try these selection procedures in practice by examining the different usage of the R(R Development Core Team, 2006) command `step` and `step.aic`.

The same data sample is reused at each step leading to overly optimistic model quality estimates (see Burnham and Anderson (2010)). The reason for this is based on the selection process - not the selected model. If the model variables, that are chosen for the best model search by the procedure, were selected independently, these models could not be influenced by the selection process. But the selection process rules determine the choice of the next model and this results in a process-bias subset for the whole model space. The fact has already been pointed out in the introductory chapter on page 10 in §1.4.1.1.

Three different kinds of bias are distinguished by Miller (1984), that is the omission of variables, the competition for selection and the stopping rule. This can lead to an extremely bad best model and there is seldom an easy way to completely bypass this problem. Exploring the whole model space, that is at least $2^n - 1$ models where n is the number of independent observations, is not a good idea either. "Let the computer find out" is a poor strategy and usually reflects the fact that the researcher did not bother to think clearly about the problem of interest and its scientific setting - according to Burnham and Anderson (2010),p. 147. By following the indexed term "data dredging" in other contexts more arguments of the same kind can be found. Unsupervised algorithms provide unpredictable results. An example is later given later (§3.3.3.1, p. 30)

Another fact that is unsatisfying about any single model selection process is, that many models are discarded by the process - whenever the stopping rule does not apply - although there will be good models among the discarded ones. Nonetheless these are dropped from consideration

☞ by the process due to some "optimal" selection criterion or because

☞ other model tests offered better results or

☞ the stopping rule did not apply.

That implies that valuable information that had cost time to create has been lost. In the single best model scenario there can be only one best model. If this best model is over-fit or lacks some crucial variables then the process was probably poorly - or worse, blindly - chosen. Cross validation (see §3.1.1) can be very helpful to avoid this situation.

Despite the problems that arise from the single best model selection approach there are many advantages, too. When the process is manually controlled or at least the results validated the selected best model offers an interpretable result based on the selected variables. Since the best model has been chosen out of a set of competitors the term best is justified by the process. The selection process is computationally cheap and relatively easy accomplished. If enough data is available for validation splitting the data is advisable. One set is used to compute the model. The remaining samples are used to validate the created model. (see again §3.1.1)

Following the ideas of Burnham and Anderson (2010) the bias from the forward or backward variable selection process can be avoided if a set of candidate models can be derived by a priori considerations. Applying scientific working methods many models can be dropped from the candidate set if a model does not conform to given scientific facts or prior knowledge. An example given by Burnham and Anderson (2010) is about cement hardening as found in chapter 3. The cement observed requires a chemical mixture of at least two different components, so any model containing only one variable must be excluded from a valid candidate set. Any technical constraint must be honored whenever models are selected. Using standard modeling software this usually means that the computed models must be manually revised. Candidate sets are to be kept clear of useless models. Violating such constraints leads to wrong conclusions - and wrong conclusions can lead to dangerous and even harmful results. If the cement hardening would be tried with the single component there would be no reaction but no harm would be caused either - but in case of medical data the wrong medicine or treatment can either be of no effect or even so harmful that this may result in death. Parsimony is an important principle that should be honored always. Too specific models reproduce the given data sample too meticulously and thus fail when generalized. Still technical constraints pose another problem that must be dealt with in some way. The above example excludes the most parsimonious models using scientific knowledge. Consequently not selection criteria alone should be considered but scientific considerations must be taken into account, too. If a model does contradict given knowledge this model is usually meaningless. An expert in this scientific field should be consulted to revalidate current knowledge versus the data set if one ore more contradicting models are found.

The term *selection criterion* used above can be understood in the greater context of *information criteria* or test based approaches as well. Using a variable selection procedure employs consecutive hypothesis tests whether to keep or drop a variable. Although frequently used in practice variable selection by hypothesis testing can end in paradoxical situations. An example is given in Burnham and Anderson (2010) chapter 3.5.5 where a hypothesis test produces an interaction term that is not supported by information based model selection. The reasoning why this test produces an unsupported interaction is quite interesting. In the given example the tested null hypothesis is false - so it should not be tested at all. Thus even a significant test result does not automatically lead to a model that should be used for inference. In chapter 3.4.6 they (Burnham and Anderson (2010)) provide another example where the model selection frequencies are compared between hypothesis based selection and information criteria based selection. They advocate and reason strongly for using only information based selection.

These *information criteria* can be utilized in weighted model averaging methods. This leads to the idea to consider more than one model for inference.

3.3 Model Ensembles

The term model ensembles is used when not a single best model is selected for inference but a group or ensemble of models is combined. Model ensembles do not suffer from model selection bias in the same way as single models and often excel in prediction accuracy. On the downside the results can seldom be interpreted straightforwardly. This section introduces multi model inference and gives an overview over recent ensemble techniques as well as their advantages and disadvantages.

3.3.1 How to gain inference from more than one model

As mentioned in (§3.2) data analysis often yields more than one good model. In the case when more than one good model exists these models are collected in a so-called *candidate set*.

A bit of formal preparation is helpful if more than one model should be taken into account. Averaging and weighted averaging for forecasting or classification is quite straightforward and utilized in practice by the *bagging* method (§3.3.2). To gain useful weights for averaging, the models can be ranked according to an information criterion and then the weights derived from this criterion:

The difference from the best model, with respect to an information criterion can be written as

$$\Delta_i = |IC - IC_{best}| \tag{3.1}$$

where IC_{best} is the best quality encountered in the candidate set. Burnham and Anderson (2010) provide empirical rules of thumb how much difference still provides support for the inferior model for AIC and related criteria. The Δ_i are used for model selection and referred to as *Akaike difference*.

Because Δ_i, the information criterion difference - or AIC difference for all subtypes of AIC criteria - depends on the relative size of the best value another selection factor would be more handy, a factor independent of the scale. AIC based selection criteria are derived from the Kullback-Leibler Information which is related to the maximum likelihood principle (see also §2.2.1.2 on p.15). Using Δ_i in the likelihood context produces

$$\mathcal{L}(g_i|x) \sim e^{-\frac{1}{2}\Delta_i} \tag{3.2}$$

Using this relation a normalized measure for the relative likelihood, the so called *Akaike weights* can be defined:

$$w_i = \frac{e^{-\frac{1}{2}\Delta_i}}{\sum\limits_{c=1}^{C} e^{-\frac{1}{2}\Delta_c}} \tag{3.3}$$

where C is the number of models in the candidate set.

Since the difference is related to the likelihood of the model and the weight is normalized to one this weight represents how likely model i is the Kullback-Leibler best model under the constraint that the best model is contained in the candidate set. Given a prior distribution the formula can be enhanced by multiplying each likelihood by the prior weight. These ideas are presented in detail in Burnham and Anderson (2010) 2.6 - 2.10.

Given the Akaike weight of a model formulae for model averaging can be defined:

- the predicted value of interest θ can be averaged by the weighted sum of all estimates

$$\bar{\theta} = \sum_{c=1}^{C} w_c \hat{\theta}_c$$

- regression coefficients β can be averaged according to these weights, too - the normalization factor is altered with respect to the existence of the coefficient in the model.

$$I_j(g_i) = \begin{cases} 1 \text{ if predictor } x_j \text{ is in the model } g_i \\ 0 \text{ else.} \end{cases}$$

$$\bar{\hat{\beta}}_j = \frac{\sum\limits_{c=1}^{C} w_c I_j(g_c)\hat{\beta}_{j,c}}{\sum\limits_{c=1}^{C} w_c I_j(g_c)}$$

Of course weighted averaging is a general principle and similar weights can be gained by bootstrapping or bayesian methods.

The next part presents practical methods where sets of models are either averaged or assessed via bootstrap or bayesian methods.

3.3.2 Bagging

Bagging, develped by Breiman (1996), is an acronym for **B**ootsrap **agg**regegation. This technique can be used with classes or numerical responses. Let $\mathcal{D}$ be a learning set for the data $\{(y_n, x_n), n = 1, \ldots, N\}$.

Choose $\mathcal{D}_k$, a sequence of learning sets that are distributed like $\mathcal{D}$ and consist of N independent observations. The $\mathcal{D}_k$ will usually be generated employing the bootstrap technique (Efron and Tibshirani, 1994).

Assume $\varphi(x, \mathcal{D})$ is a predictor for y from the input x.

To produce a better predictor than the single learning set predictor $\varphi(x, \mathcal{D})$ bagging aggregates the results from the learning sets $\mathcal{D}_k$.

If the response is numerical the aggregated result is averaged, in the case of discrete classes the majority vote is used.

$$\varphi_b = \begin{cases} avg\varphi(x, \mathcal{D}) \text{ if the response is numerical} \\ argmax_j nr\{k | \varphi(x, \mathcal{D}) = j\} \text{ otherwise} \end{cases}$$

According to Breiman (1996) bagging will produce more accurate prediction if the single method is unstable. That means that small changes in the data result in different predictions. If a selection method is stable, like k nearest neighbors - bagging fails to improve the prediction accuracy. Worse, bagging might end up with an inferior predictor especially if the single predictor is poor.

Another aspect of bagging is that the result cannot be interpreted easily. The aggregation in the numerical case will result in a weighted vector of the single $\mathcal{D}_k$ results. This vector is very hard to interpret: Each single result can be interpreted the usual way and the weight is an overall goodness of fit in a candidate set. Yet the averaged result does not preserve a familiar structure. If no interpretation of the result is necessary, bagging can provide prediction with increased accuracy. The precondition of instability should be met, though. For example bagging neural nets can produce nice results, see Sohn and Dagli (2003).

3.3.3 Boosting

Another variant of model ensembles is the boosting algorithm, devised by Schapire (2008).

Basically boosting is a procedure that improves (boosts) the accuracy of a *learning algorithm*. A "weak" learning algorithm that just performs better than random guessing can be boosted into an arbitrarily "strong" learning algorithm (see Schapire (1990)). Combining the majority votes

over weighted sets of predictions produces a superior prediction method.
The original boosting will process a data set $\mathcal{D}$ with n samples the following way:

1. select a subset $\mathcal{D}_1$ from $\mathcal{D}$ where $n_1 < n$ and train a weak learner $\mathcal{C}_1$ with this subset $\mathcal{D}_1$.

2. select a subset $\mathcal{D}_2$ from $\mathcal{D}$ where $n_2 < n$ which consists of half of the samples that were misclassified by $\mathcal{C}_1$ and train a weak learner $\mathcal{C}_2$ with this subset $\mathcal{D}_2$.

3. select all samples from $\mathcal{D}$ where $\mathcal{C}_1$ and $\mathcal{C}_2$ produce different results and train the weak learner $\mathcal{C}_3$

4. the final classifier will be the majority vote of all the weak classifiers - similar to bagging at this point.

The algorithm AdaBoost developed by Yoav Freund and Robert Schapire employs re-weighting rather than resampling the samples for the learner:

1. Start with uniform weights $w_i = \frac{1}{n}, i = 1 \ldots n$

2. For all $m = 1 \ldots M$ do

   2.1. fit $f_m(x) = \{-1, 1\}$ using weight w_i from the training data

   2.2. $\varepsilon_m = E_W \left[1_{(y \neq f_m(x))} \right]$ $c_m = \ln \frac{1 - \varepsilon_m}{\varepsilon_m}$

   2.3. update $w_i^m = w_i^{(m-1)} e^{c_m \cdot 1_{(y \neq f_m(x))}}$ and rescale $\sum_i w_i^m = 1$

3. output classifier $sign \sum_{m=1}^{M} c_m f_m(x)$

Today many variants of the boosting algorithm have emerged. To name a few the *Adaboost** (Rätsch and Warmuth, 2005), WeightBoost (Jin et al., 2004), BrownBoost (Freund, 2001), LogitBoost, Gentle AdaBoost, Real AdaBoost (Friedman et al., 1998), FloatBoost (Li et al., 2003), KLBoost (Liu and Shum, 2003), DiscreteAdaBoost (P. Viola, 2002).

Hastie et al. (2003), chapter 10 offers a nice introduction in boosting and provides some examples where boosting is used on regression trees.

3.3.3.1 Example: Boosting a regression model

Boosting can be useful in a more general context than classification. As boosting is a quite recent technique, an example showing how boosting is used on regression models is presented here. For this example the additional R(R Development Core Team, 2006) package **mboost** (Hothorn and Buhlmann, 2007) is used. The hills data set (Atkinson, 1986b) as introduced in figure 1.4 on page 5 will be fitted, this time by non-linear models. The basic idea is that since some of the observations have fairly high leverage and boosting has been developed to cope with problematic observations we might find a nicer model than the plain linear model.

As a reference goodness of fit $R^2 = 0.8456$ and $AIC = 312.8124$ can be used. The initial model *time* $\sim$ *dist* is the source for these measures. This model was chosen because it provides the best fit from all two-dimensional models and to facilitate presentation only two dimensions will be considered.

After loading the *mboost* library and the *hills* data we fit as many bootstrap samples as we have samples in the data set with the gamboost-function:

```
> hills.gb<-gamboost(time~dist, data=hills, dfbase = 4,
+ control = boost_control(mstop = max(dim(hills))))
> hills.gb

        Generalized Additive Models Fitted via Gradient Boosting

Call:
gamboost.formula(formula = time ~ dist, data = hills, dfbase = 4,
+ control = boost_control(mstop = max(dim(hills))))


        Squared Error (Regression)

Loss function: (y - f)^2


Number of boosting iterations: mstop = 35
Step size:   0.1
Offset:   57.87571
Degree of freedom:   4

> AIC(hills.gb)
[1] 6.879241
Optimal number of boosting iterations: 35
Degrees of freedom (for mstop = 35): 5.003767
```

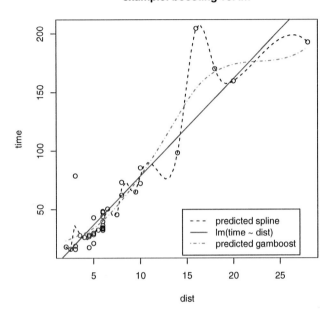

Figure 3.1: Scottish hill races: boosted model, spline and linear model

Fitting a non-linear spline model is definitely too complex for these data. The remaining degrees of freedom suggest that this model has been over-fitted. Compared to a smoothing spline

```
lines(smooth.spline(hills$dist, hills$time), lty=3, col = "blue")
```

the result looks very nice, though. The spline offers another model that is even over-fitted worse with respect to the observed data. Nevertheless this boosted model has been found too complex though not over-fitted to such an degree as the spline.

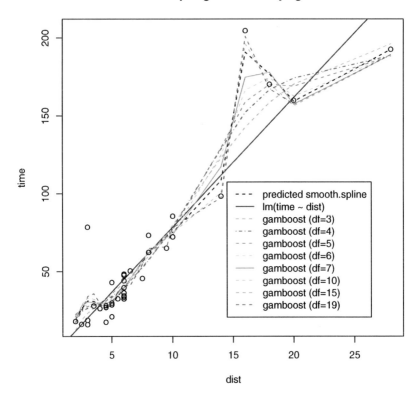

Figure 3.2: Scottish hill races: different gamboost models, varied in complexity, in comparison to the reference linear model and spline

The concept of complexity from *dfbase* in terms of degrees of freedom has to be interpreted in another way as in the usual model context. Figure 3.2 shows that the higher the *df* is chosen the more specific is the gamboost model. The cause is that *dfbase* fixes the degree of the smoothing spline.

Consequently there must be an optimal *dfbase* that will yield an AIC-best boosted model:

dfbase	df	AIC
2	1.950080	7.121613
3	3.754981	6.900033
4	5.003767	6.879241
5	6.290985	6.869196
6	7.571787	6.822259
7	**8.812952**	**6.788193**
8	9.717647	6.789715
9	7.571787	6.827064
10	10.47516	6.902196
15	14.56801	7.564965
19	16.43338	8.01779

Table 3.1: gamboost complexity and quality

Choosing an optimal boosted model from the above by *AIC* picks the gamboost model with *dfbase* = 7 that features a better quality and less complexity than the initial (and by default) chosen *dfbase* = 4. See figure 3.3.

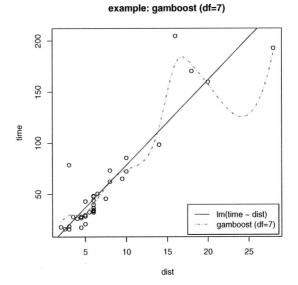

Figure 3.3: Scottish hills data set: the AIC-best gamboost model

Considering the comment from Atkinson Atkinson (1986a) on Chatterjee and Hadi (1986) about treating high leverage points in the hills data set it is unwise to fit such a non-linear model. Outliers are to be identified and removed before fitting models. The paper discourages the use of transformations in the presence of high leverage points. Fitting a non-linear model for the data without removing the outliers is similar to transforming data non-linearly first and fitting the model afterwards. A priori considerations forbid such a complex model too, because physically time and distance relate in a simple functional way to each other. The function represented from the best boosting model suffers from inexplicable non-linear effects where *dist* is greater than 15. Possibly human athletes suffer from dwindling power after a certain effort. An explanation, why the the level of exhaustion declines after that point and the overall performance improves afterward is hard to imagine, though. The steep level of "magic" regeneration shows that this model is not fit for inference.

This example demonstrates that the best algorithm will produce a poor result when used mindlessly. Boosting is used to fit hard-to-fit observations, so outliers must pose problems. Heeding the advise from Atkinson (1986a) to omit the observations 7,18 and 33 and rerunning the above example with the reduced data set chooses a simpler model :

dfbase	df	AIC
2	1.944510	6.487553
3	**3.717086**	**6.411345**
4	4.904073	6.427204
5	6.03122	6.473055
6	7.571787	6.822259
7	7.096864	6.527251
10	10.63354	6.877117

Table 3.2: gamboost complexity and quality (outliers removed)

AIC minimum chooses the model with *dfbase* = 3 that will be much closer to the original line (see 3.4). The best boosting model is now very close to the linear model. This slight downward bend might be explained by human exhaustion, this could also be used as a model despite the extra complexity.

This seems like a feasible explanation, nevertheless the fact should be mentioned that the *gamboost* model is contained in the 95% confidence area of the original linear model.

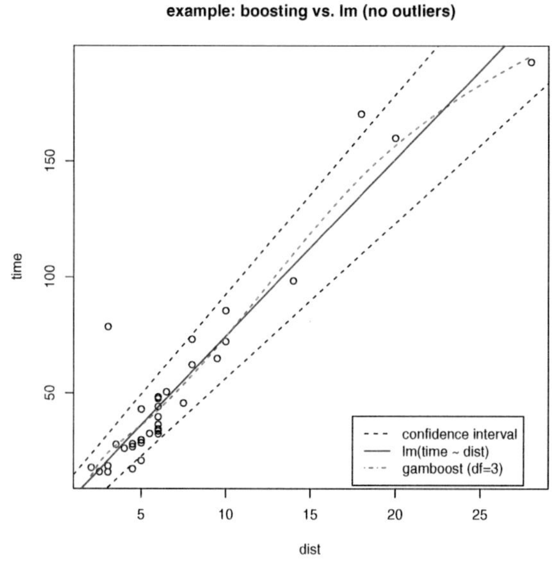

example: boosting vs. lm (no outliers)

Figure 3.4: Scottish hill races: the AIC-best model where outliers have been removed. Contains confidence interval of reference linear reference model.

As pointed out boosting is a very powerful method to fit models to data-sets that contain hard-to-fit observations. Outliers skew the result and the boosting algorithm over-fits. Consequently data cleaning must precede model fitting, especially when boosting is applied.

3.3.4 Bumping

Bumping is an acronym for **B**ootstrap **u**mbrella of **m**odel **p**arameters. One of the most annoying shortcomings of bagging and boosting is the loss of interpretability. Another way to utilize the bootstrap method for model averaging that tries to improve prediction accuracy but still remain interpretable is bumping (Tibshirani and Knight, 1995):

1. Start with a training sample $\mathcal{D}$ that is independently and identically distributed from a distribution $\mathcal{F}$. The model for the data depends on the parameters θ that is chosen according to an *target information criterion* $\mathcal{R}(\mathcal{D}, \theta)$.

2. Calculate $\hat{\theta} = argmin_{\theta}\mathcal{R}(\mathcal{D}, \theta)$

3. Assume we have a competing information criterion $\mathcal{R}_0(\mathcal{D}, \theta)$ the *working criterion*[2]

---

[2]The *working criterion* is not necessarily different from the *target criterion*

4. $\hat{\theta}$ will be estimated by drawing bootstrap samples $\mathcal{D}_k, k = 1..m$, resulting in a training estimator $\hat{\theta}_k = argmin_\theta \mathcal{R}_0(\mathcal{D}_k, \theta)$

5. The *bumped estimate* of θ is $\hat{\theta}^B = argmin_\theta \mathcal{R}(\mathcal{D}_k, \hat{\theta}_k)$

- that improves robustness against outliers and

- has a better chance to find the global minimum, and additionally

- since a parameter set is selected and not aggregated the underlying structure is preserved and the result remains interpretable.

Bumping often improves the performance of the model estimate. Robustness against outliers is improved, because bootstrap selection will leave out outliers on occasional sub-samples. It can simplify the optimization procedure, especially in the case of constrained optimization and the chances of finding a better solution than a local minimum are enhanced.

3.3.5 Random Subspace Method (RSM)

This method will train learning machines on randomly selected subspaces of the original data set. The subspaces are selected in a way that from an originally n dimensional feature space an r dimensional subspace is selected where $r < n$. The randomization is utilized to pick different combinations of variables for further processing.

The output is aggregated in a second step. Ho (1998) gives an example how this methods improves selection accuracy on classification trees while keeping the complexity at a reasonable level. A comparison of boosting, bagging and the random subspace method is given in Skurichina and Duin (2002) and an application on a multi-feature search in Lai et al. (2005).

3.3.6 Bayesian Model Averaging

Bayesian Model Averaging leverages the Bayesian posterior distribution as a means to combine models. Given the data $\mathcal{D}$ and models $\mathcal{M}_1 \ldots \mathcal{M}_n$ these posterior distribution for a model $\mathcal{M}_k$ can be written as

$$P(M_k|\mathcal{D}) = \frac{P(\mathcal{D}|M_k) \cdot P(M_k)}{\sum\limits_{i=1}^{K} P(\mathcal{D}|M_i) \cdot P(M_i)}.$$

where

$$P(\mathcal{D}|M_k) = \int P(\mathcal{D}|\theta_k, M_k) \cdot P(\theta_k|\mathcal{M}_k) \partial\theta_k$$

with θ_k the vector of parameters for model $\mathcal{M}_k$, $P(\mathcal{D}|\theta_k, M_k)$ the likelihood and $P(\theta_k|\mathcal{M}_k)$ the prior density.

In practice the form of the prior distribution is rarely known in advance so many forms of a feasible prior distribution have been discussed (see Fernandez et al. (2001), Yin and Davidson (2004), Lahiri (2002), Hoeting et al. (1999)). If there is no prior knowledge a uniform prior $P(\mathcal{M}_k) = \frac{1}{n}$ or a prior penalized according to the model complexity can be used initially (Yin and Davidson, 2004).

Stepwise this previous knowledge - the initial distribution - is adjusted by an optimization algorithm. Finally the posterior distributions approximates the distribution how likely a model is chosen. Either a subset from the initial models is selected via a threshold probability and used for averaging. Otherwise, if the model set is small enough to take advantage of previous knowledge or other constraints - the whole set can be considered for averaging.

The averaging procedure will differ according to the modeling goal. If the goal is variable selection the importance of the variables can be used straightforwardly by averaging the model variables from the posterior distribution. An example is given in the appendix (see A.1.1 on p. 145). Another useful application is ensemble selection or forecasting. A practical example how to combine model ensembles using bayesian model averaging can be found in Raftery et al. (2003).

3.3.7 Stacked Generalization

Stacked generalization, developed by Wolpert (1992), also known as stacking, leverages predictions by combining different models. An estimator or classifier is perceived as a *generalizer* that guesses a parent function based on a set of sample mappings $\mathcal{D}$ the learning set. The term *generalizer* is chosen because the main purpose of stacking is to reduce model bias and thus upgrade the prediction accuracy on unknown samples. Stacking means that these generalizers can be linked in the way that the output of one generalizer is fed to the next level generalizer. This stacked generalizer can be interpreted as a self-contained generalizer as well.

Generalizer has been defined by Wolpert (1992) as a mapping from a learning set of m pairs $\{x_k \in \mathbb{R}^n, y_k \in \mathbb{R}\}, 1 \leq k \leq m$ together with a question $\in \mathbb{R}^n$ into a guess $\in \mathbb{R}$. Since stacking can work with an arbitrary generalizer - according to the above definition - models of different types can be combined. The algorithm can be understood as a smarter brand of cross-validation starting with distinct splits from the original data sample, where the generalizer is trained on one subset but the quality is measured on the remaining part.

There are some technical issues on how to combine generalizers for stacking, see Ting and Witten (1997) and Ting and Witten (1999). But the method has been used successfully, too according to Ghorbani and Owrangh (2001).

3.4 Summary

Ensemble methods outperform the single best model approach in many ways. Since more then a single best model is used ensembles do not suffer from bias in the same way. Stacking is especially designed to reduce model-induced bias with its main focus on prediction accuracy improvement for unknown data. Boosting tries to focus on hard-to-fit points and thus is more susceptible to over-fitting. Bumping tries to cope with outliers by finding a best-fitting model among bootstrapped competitors. BMA produces another weighting pattern for model averaging purposes and can be used for many selection type problems too.

All of these procedures have in common that not a single model is consulted for inference but many models. Each model - if a correct selection procedure has been chosen - might be considered for inference on its own. Candidate sets collect all these good models and try to extract the combined knowledge. Hence good candidate sets for ensembles need to be assembled by hand. Many practitioners shy away from this effort. Automatic bootstrap-driven procedures are offered by lots of software packages today, but handling candidate sets is still very time-consuming.

Still the improved prediction accuracy, reduced model bias and robustness against outliers are good reasons to construct candidate model sets.

Chapter 4

Interactive Model Analysis

Various ways for producing models or model ensembles have been presented in the previous chapters. These procedures are not meant to run unsupervised though it is tempting to fit a sophisticated model straight away. §3.3.3 (p. 28) gives an example that this approach leads to wrong results. Boosting is a powerful tool to fit a model along hard-to-fit points. High leverage points are likely to deform a simple model into a complex over-fitted one. The used data set includes outliers and these outliers were fit like ordinary high leverage points. The fitted model was too specific as the model was twisted to provide an optimal match for these outliers. After the outliers were removed a reasonable model could be found. This fact is repeated so many times to emphasize the importance of data cleaning. Outliers skew models and require careful treatment - even elimination is an option.

For the task of outlier identification diagnostic statistics like Cook's distance (Cook, 1977) can be employed to compute the influence of a single observation. More specifically the difference between observation and forecast of regression model is used. This distance statistic derived from each residual provides a useful diagnostic plot. Revisiting the introductory example §1.3.1 (p. 5) the static Cook's distance, which measures the leverage for the models, 1.5a and 1.5b (p. 7) shows the following:

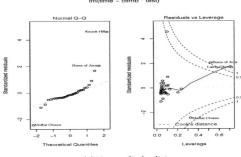

(a) $time \sim climb \star dist$

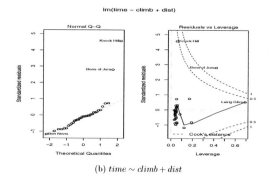

(b) $time \sim climb + dist$

Figure 4.0: Scottish hill races data set, QQ- and leverage plot for two linear models.

The figures 4.1a and 4.0b exhibit two different types of plots. On the left a quantile-quantile-plot (qq-plot) is used to validate distribution assumptions. Since a linear model is fit the error term distribution is expected to be Gaussian. Some observations are named to identify those that require further investigations. Remembering Atkinson (1986a) the observations 7,18 and 33 were identified as outliers. The matching names are "Bens of Jura", "Knock Hill" and "Two Breweries". The first two outliers are named in both qq-plots. That's one hint so far to remove these observations. Another hint is found in the right plot. The residuals are plotted versus the leverage and two lines identify potential outliers by displaying Cook's distance. As a rule of thumb a Cook's distance higher than 1 requires validation of the observation. By this rule "Bens of Jura" was identified as an outlier on two occasions.

Unfortunately not every outlier is simply identified by some static plots. Each possible outlier must be located repetitive after refitting the model, e.g by revalidating distribution and leverage. The steps in between are left out but the outliers were identified using the method as described above:

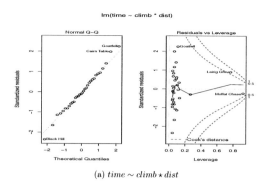

(a) $time \sim climb \star dist$

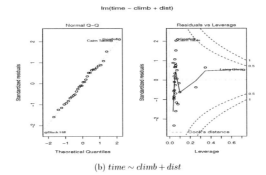

(b) $time \sim climb + dist$

Figure 4.0: Scottish hill races data set where outliers have been removed, QQ- and leverage plot for two linear models.

After three steps no observation remains, that features a critical leverage, so 32 out of 35 observations are considered fit for modeling. Consequently the results of the introductory models (§1.3.1, p. 5) fit to uncleaned data should be discarded and new models fit to the revised 32 observations. Nonetheless this example demonstrates basic concepts and pitfalls.

The data cleaning step was left out in the introduction to show a basic concept and to keep the example concise. In fact step by step examination, cleaning up the data set and refitting models, is laborious:

```
> # check the distribution for hills using a qq-plot
> plot(lm(time~climb+dist,data=hills[-c(7),]),which=2)
> plot(lm(time~climb+dist,data=hills[-c(7,18),]),which=2)
> plot(lm(time~climb+dist,data=hills[-c(7,18,33),]),which=2)
> plot(lm(time~climb*dist,data=hills[-c(7),]),which=2)
> plot(lm(time~climb*dist,data=hills[-c(7,18),]),which=2)
> plot(lm(time~climb*dist,data=hills[-c(7,18,33),]),which=2)
```

⋮

Given a practical high dimensional data set this involves enormous efforts. For the task above a tool would be nice to select, or remove, observations stepwise and monitor the impact at once.

4.1 Interaction

The problem above involves model fitting for each selected subset. Iplots (Urbanek and Wichtrey, 2003) - a package that is specialized in interactive graphics - is able to handle selection events and refit models interactively. The observations might be selected from any (i-)plot; for 35 observations a scatter plot is a feasible source:

```
iplot(x=dist,y=time)->ref
iabline(lm(time~dist),plot=ref)
iplot.set(1)
```

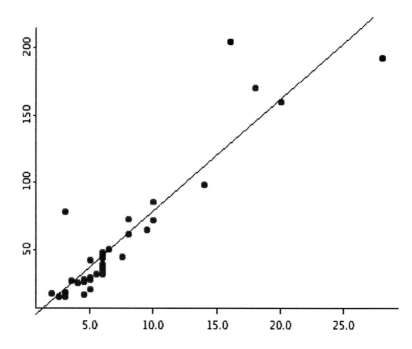

Figure 4.1: interactive selection using iPlots. iPlots offers various interactive features, like automatic model fitting on a selection from the data set. (shown here)

Using iplots event mechanism the selected data subset will now refit models interactively whenever this subset is changed:

```
while (!is.null(ievent.wait())) {
    if (iset.sel.changed()) {
        s <- iset.selected()
        if (length(s)>0) {
            par(mfrow=c(1,2))
            plot(lm(time[s]~climb[s]+dist[s]),which=2)
            plot(lm(time[s]~climb[s]+dist[s]),which=5)
        }
    }
}
```

The commands above produce the same type of plot as (Figure 4.0b). Instead of taking lots of steps in the form

- select subset
- plot model

the subset selection part triggers a refitting of the model and the plot is updated simultaneously. Additionally the subset is now effectively selected by mouse on this plot. This combination saves a lot of time and work compared to typing in the model command for each subset manually. This example demonstrates the benefits of *interaction*. Yet no formal definition for the term interaction has been provided. *Interaction*, in the context of human computer interfaces is required to meet some requirements. Interactive statistical software running on a computer

☛ must run fast, so the user is not forced to wait for results.

☛ the graphical user interface is easy to use.

☛ interrelation between variables can be visualized

by selecting observations and

watching the immediate response, usually highlighting, in related plots.

☛ :

This list contains some features instead of giving a clear definition. *Interaction* literally means that two or more objects affect each other. In the context of statistical software these objects are data, observations and derived objects like models, forecasts and residuals. Most interactive software packages provide means to select objects on plots while other plots containing related data are simultaneously affected by this selection.

Interactive plots are a great improvement to step by step plotting. To explore a practical sized data set without interactive aid is unthinkable today. Exploratory data analysis (EDA), as proposed by Tukey (1977) benefits from interactive techniques, too.

EDA is an approach for data analysis and employs various techniques - lots of graphical ones - to gain maximum insight into a data set. To achieve that goal model assumptions are delayed as long as possible. A head-on modeling approach seldom leads to precise models, as the boosting example (§3.3.3, p. 28) demonstrated.

Naturally the importance of interactive means grows with increasing data set size. The reason is that the effort to separate outliers from underlying structure grows with size. Even selection becomes more difficult when the data is crowded. Different helpful selection methods have evolved:

selection by simply clicking on a point of a plot or a reference on a spread-sheet grid. Usually a single observation is selected using one click, to select more than one observation at a time either another key needs to be pressed or one of the following is used:

brushing instead of repeated clicking the button is held down. Until the button is released all points that have been brushed are selected.

shape based selection by holding down the mouse a rectangular [1] area is outlined where all inner points are selected.

---

[1] circular or elliptical shapes are rarely used

lassoing like a lasso an outer frame is drawn around a table or plot and all surrounded elements are selected

zooming changing the scale on a plot while scrolling through smaller partitions of the data gives deeper insight where data gets crowded

focussing when lots of data is available only a subset is viewed at a time. This focussed subset is used for later depending plots.

weighting many types of two-dimensional plots are able to handle a third dimension. For example, a scatter plot with additional weight dimension is called a *bubble plot* (see e.g. figure 5.5, p. 58).

variation of display many plots provide different options how they are displayed, like the number of bins used in a histogram. Many plots support the variation of the scale, e.g. linear, logarithmic or a custom variation.

Other methods do not affect the selection but leverage knowledge about single observations or groups or sets.

querying can be interpreted as a backward link from the graphical representation to the data. Using histograms a query can show the exact number of cases contained in a bin. Additionally, if this number is not too large, an identifier for each observation can be shown - otherwise quantile information or anther summary type of information can be given. Querying is used on a single plot and does not affect the *selection*.

linking basically means that a connection between data and plots exist. These links can be logical, an observation of many dimensions for instance, or created by selection or grouping. For graphical operations *linking* means that selected items are simultaneously highlighted in different views. A view can be a simple list, a table or any graphical plot. If the views are linked the *selection* is synchronized among each of these views, so each change can be tracked on many dimensions.

Each method mentioned above is designed to be used for data, and especially in combination with graphical plots. In the context of multi-model management the amount of data explodes, in comparison to the original data.

Each model contains model specific data and either the forecast or residuals are of interest. For example parallel coordinate plots provides one way to draw conclusions from residuals. Furthermore the model specific data is of interest. To describe the structure of a data set a wide range of different models can be considered. Model averaging techniques provide weighted mixtures; so in a way an averaged multi-model behaves as one model. But averaging hides many effects that can be observed in single models or even groups of models.

4.1.1 An example

Using all 35 observation from the hills data set §1.4 (p. 5) 50 bootstrap samples are drawn to calculate the models

1. $m1 = time \sim dist$

2. $m2 = time \sim climb$

3. $m3 = time \sim dist + climb$

4. $m4 = time \sim dist + climb + dist : climb$

The full data set - including the identified outliers - was chosen by design, since the outliers will provide a wider range of values than the cleaned up subset would. 200 models were fit and the following number of coefficients are available:

(intercept)	dist	climb	dist:climb
200	150	150	50

Table 4.1: Numbers of coefficients for 50 bootstrapped models

Comparison   The outliers and the randomized subset selection affect the model parameters. In this first step the coefficient values are compared for the cleaned data set, the full data set and average and range for the bootstrap samples (bs):

formula	cleaned	full	mean bs	min bs	max bs
time ~ climb					
(Intercept)	12.262	12.699	13.070	2.405	27.860
climb	0.023	0.025	0.025	0.016	0.031
time ~ dist					
(Intercept)	-3.725	-4.841	-7.261	-33.400	9.958
dist	7.467	8.331	8.731	5.323	12.720
time ~ climb + dist					
(Intercept)	-8.230	-8.992	-7.364	-17.640	3.754
climb	0.007	0.011	0.010	0.004	0.016
dist	6.635	6.218	6.256	3.274	7.215
*time ~ climb * dist*					
(Intercept)	-6.466	9.395	10.300	-13.810	33.510
climb	0.006	-0.001	0.001	-0.011	0.010
dist	6.390	4.149	3.911	-0.014	7.246
climb:dist	0.000	0.001	0.001	-0.000	0.002

Table 4.2: parameters for subsets/bootstrap samples of hills

Even such a small example provides an amount of data that is too large to evaluate without statistical plots. An interactive table like a spread sheet program gives no more insight than textual summaries. A summary provides quartile information for each coefficient value; this value consists of a p-value and other test statistics. This is less information at one view than a box-and-whisker plot offers. Table 4.2 displays information about the model parameters using different subsets of the hills data set. "Cleaned" refers to the reduced data set without the identified outliers, "full" contains all 35 observations while the mean, min and max from the bootstrapped samples are shown. Figure 4.2 displays the p-values and values of the model parameters grouped by model.

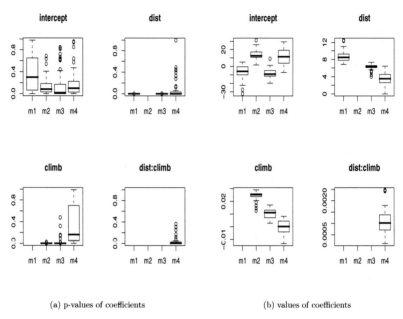

(a) p-values of coefficients   (b) values of coefficients

Figure 4.2: Scottish hills data set: model coefficient p-values and values for the 50
bootstrapped models

The range is narrow considering the used data set is a text book example for high leverage
point treatment, and all high leverage points were available for sampling. While p-values for
non-intercept terms are bounded by a narrow interval the intercept varies greatly among the
different samples gained via bootstrapping.

To find out more about one special model a static plot is no longer practicable. Since all outliers
were available for bootstrapping some peculiar parameters should be analyzed. For this quick
example the parameter with the highest *dist* p-value is chosen.

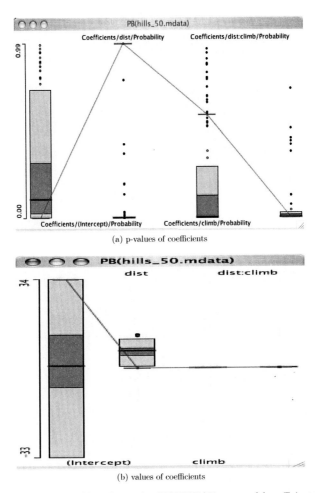

(a) p-values of coefficients

(b) values of coefficients

Figure 4.3: interactive model analysis using MONDRIAN, one model coefficient that has a high p-value is analyzed

Figure 4.3 covers all four types of models, so a next step is to find out what observations were used for that model. Using the database directly would be one option, querying the best choice. Model type m4 is identified from the plot, as all 4 parameters are present: $time \sim dist + climb + dist : climb$. Querying the selection a model is identified as #259 (the # indicates an internal database identificator). Alas Mondrian does not know more about the data set, or the database so the bootstrap sample can only by identified using MORET. Remembering the outliers were identified by the numbers 7, 18 and 33. Model #259 contains 3 instances of the 7th and 2 instances of the 33rd observation. The data set, more precisely two peculiar bootstrapped subsets, illustrate the bias induced by augmented outliers. In the weighted scatterplot (Figure 4.4) the number of sub-sampled observations can easily be seen:

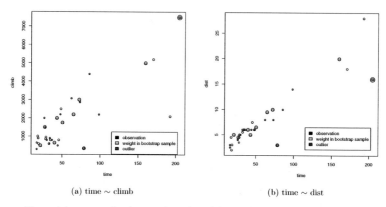

(a) time ∼ climb   (b) time ∼ dist

Figure 4.4: scatterplot for two skewed models, weighted by bootstrap frequency

This is one way to revalidate models, and discover that outliers were involved. Of course fitting models the other way round is faster and more effective:

1. clean up data

2. analyze data

3. fit one or many models

Averaging Averaging is straightforward. As long as no other weighting schema is known all parameters are equally weighted. Later Akaike weights will be introduced; the Bayesian approach is another well-known weighting schema. This results in

Variable/estimate	(intercept)	climb	dist	climb:dist
Averaged	2.188	0.11	6.299	0.001
m4, 32 Observations	-6.466	0.006	6.390	0.000
m4, 35 Observations	9.395	-0.001	4.149	0.000

Table 4.3: Comparison of parameter estimates - averaged, complete data set and cleaned data set

The averaged estimates of the bootstrap models look conclusive. So the averaging effect overrules most of the bad influences - effects from outliers - that are hidden in the single models. Using the interactive microscope some obscure patterns can be identified. Any model fit on data containing outliers does not feature good generalization properties. Therefore high leverage points and outliers must be identified and treated first.

4.1.2 Summary

That closes the circle to EDA-techniques. A data set is the source for model analysis. When a model set is involved the amount of data grows. Working in such a manner as proposed by the EDA philosophy avoids loopbacks. Exceptional observation are evaluated and fit by any model. Since so much data is produced in this process *interactive* tools are a valuable asset. Many software packages support interactive analysis, though usually only one data frame can be analyzed at a time (e.g. Theus (2002b)). An example has been given how to leverage interactive graphics to validate conspicuous parameter forecasts.

Interactive methods have been briefly introduced. Theus (1996) provides a general introduction for the use of interactive tools. Unwin et al. (2006) offers many techniques how to deal with large data sets and also covers lots of interactive techniques. Urbanek (2006) employs interactive means (e.g iPlots Urbanek and Wichtrey (2003)) for the analysis and visualization of classification and regression trees.

Chapter 5

The Software MORET

The principal part of this work covers the creation of a software to manage statistical models with special regard to candidate sets. Accordingly this chapter and the following cover the largest part of this work, starting with a theoretical introduction.

First general thoughts about software development, with special regard to statistical solutions, are presented. This introduces general software development issues and philosophical thoughts. What must be created anew is matched against existing solutions. Interfaces are sketched in order to leverage established software packages. By doing so the features for this special software solution are introduced and references to other useful projects given.

Following up is a classical example analyzing an election data set.

After that the technical requirements are presented that finish with the first working version of MORET.

The historic development of MORET is presented on purpose. Many lessons can be learned from the failures of early versions - in particular when the later solution is at hand. This first version still faced many problems, especially how to cope with the large variety of model types available today. For that reason the theoretical foundation is laid out how new model types can be adapted.

At the end of this chapter other useful tools are presented that facilitate model handling. Practical examples are found in the next chapter.

5.1 General Thoughts

Before software is written an idea must be defined what that piece of software is used for. Since ideas are abstract concepts, they need to be transformed into more detailed sketches. This draft is used to create a software prototype.

At the prototype stage a lot of tasks are unclear; technical details cannot be foreseen while many useful ideas emerge during the course of development.

This sketching stage precedes an assessment of currently available software packages. No benefit can be gained when an available software provides all functions required. *Managing models* is an abstract idea, and many software artifacts were already available when the first sketches were made. In fact the list of statistical software grows longer day after day An exhaustive list of available commercial software or all R(R Development Core Team, 2006) packages will not be given. More than 2000 R packages are located in official repositories. From these more than hundred are concerned with modeling. An objective for this new software package must be specified and the distinction from available solutions has to be clarified:

✔ Most software packages provide means to calculate models, but what about *candidate sets* and *model management*.

✔ Detailed questions are required to draw a sketch for the last two aspects:

☞ What part of a model is interesting and *manageable*?

☞ What type of tasks are preformed by hand today and what can be done automatically by software ?

☞ Given models are managed by software, what needs to be done to *manage candidate sets*

As presented in earlier chapters information criteria (§2.2.2 on page 16) are one subject of interest. Information criteria are global level model summaries (§1.3.2 on page 8) but more detailed information is useful (§1.3.3 on page 8), too. Model residuals are widely used but storing all residuals for each model produces large amounts of data.

So far no other software supports the management of models in a way that candidate sets can be administered by software rather then by hand. But writing new software for model computation would is superfluous.

R(R Development Core Team, 2006) is very versatile at coping with mathematical optimization algorithms and the calculation of different model types. Utilizing such a software has many advantages over rewriting. On the one hand any user needs time and effort to familiarize with new software. Rewriting forces the user to re-learn tasks and of course rewriting is prone to new errors. Taking these points into account the foundation of model calculation will be left to R. Beside pure mathematical calculation R is extensible, so new model types can be integrated, too. Analytical plots are available and even interactive plots are supported, e.g. iPlots by Urbanek and Wichtrey (2003).

Despite R provides all tools for low level computation and plotting, extending a given R-graphical user interface (*gui*) is very hard to do. Data management requires lots of tasks a *gui* is indispensable. Another great benefit that this *gui* should provide is an ability to externalize the managed data for other interactive tools. For candidate set management either a spreadsheet program is helpful or specialized interactive plotting tools.

Generally good software should be re-used whenever possible. Only as much as required - and unavailable - should be created. An interface to use all these neat programs is favorable to new all-in-one software.

So far the above only scratched the surface of model administration. Candidate set management will be dealt with later in this chapter. To emphasize how much work manual candidate set selection involves, an example is given in detail now:

5.2 Example: Elections04

5.2.1 About The Data

Figure 5.1: Do you trust an electronic voting device? [1]

Many elections in the United States are won by a narrow margin. In this tough competition it is crucial for each candidate to know how to use his best efforts during the campaign. That is common for every democratic election system, still there are some unique aspects about the electoral system used in the USA.

To win the election not the candidate who has been awarded the most votes will be selected president but rather the candidate is proclaimed winner who gets the majority vote from the electors. Each state has a number of electors that will typically promise to support one presidential candidate and cast their secret vote before the United States Electoral College. 538 electors determine the election winner by majority vote. This setup is based on the historical fact that the USA is a country that covers a huge spatial area and collecting all votes to one point was an unthinkable thing to do in the early years of this system. So rather than bring all votes to one point each state was assigned a number of heads that would go and carry over the votes to a central point. Due to the fact that this is not equivalent to the total majority Al Gore would have been elected president by the majority of American citizens in 2000. But he lost because the majority of electors voted for George W. Bush. The millennium election has been studied in detail by Weisberg and Wilcox (2003). Lots of different ideas are presented to determine what factors led to this result. Apart from demographic influence (indirect) partisanship, economic reasons and other patterns are examined. Connections between the party of the last president, topics of election campaign, and many more - a large variety of data has been analyzed to provide a meaningful model. Despite ex-post analysis benefits from knowledge of the outcome - compared to forecasting - the explanations were seldom entirely convincing.

Regardless what any model might predict, to win the election the candidate has to win over the electors in key states rather than the majority over all votes. The term *swing state* is also commonly used for these states that are not loyal to one of the major parties over many years but tend to elect their candidate according to the election campaign. From this background the 2004 elections were monitored very closely and the term fraud has been used very quickly - but data to support evidence has to be looked upon very carefully.

The subsequent example data set has been explored in the hope to find out if the electronic touch screen voting device affects the election results. The following variables are available:

variable name	type	description
fl	categorical 0 or 1	1 if Florida
oh[2]	categorical 0 or 1	1 if Ohio

---

[2]unused

variable name	type	description
county	Text	the county
d1996	numeric	votes for Robert J. Dole in 1996
c1996	numeric	votes for Bill Clinton in 1996
b2000	numeric	votes for George W. Bush in 2000
g2000	numeric	votes for Al Gore in 2000
b2004	numeric	votes for George W. Bush in 2004
k2004	numeric	votes for John Kerry in 2004
etouch	binary	voting machine: 1 = electronic touch screen, 0 = Op-Scan
income	numeric	median income
hispanic	numeric	percentage of hispanic populations
$b00pc$ [3]	numeric	votes for Bush 2000 in percent
$b04pc$ [3]	numeric	votes for Bush 2004 in percent
b_{change} [3]	numeric	$b04pc - b00pc$
$b00pc_{sq}$ [3]	numeric	$\sqrt{b04pc - b00pc}$
size	numeric	$votes04$
$b00pc_e$ [3]	numeric	$etouch \cdot b00pc$
$b00pcsq_e$ [3]	numeric	$b00pc_e^2$
$votes04$ [3]	numeric	number of votes posted in 2004 $b2004 + k2004$
$votes00$ [3]	numeric	number of votes posted in 2000 $b2000 + g2000$
v_{change} [3]	numeric	$votes04 - votes00$
$d96pc$ [3]	numeric	$\frac{d1996}{d1996+c1996}$

Table 5.1: Variable description for the elections data set

Note that the data for *ohio* is not complete: only data for 2000 and 2004 is available but 1996 is missing. Because of that this variable will not be used for modeling here. Still *ohio* might be useful for model comparison or other purposes, so it is mentioned for the sake of completeness.

Since the effect of the variables is the primary concern it makes little sense to include all transformed version of a variable in one model. Now heeding the aforementioned, first the data set is analyzed and after that models are fit. The geographic distribution of the underived data provides an overview; using standard linear models an *etouch* effect is not easy to identify. The other variables are correlated which complicates the separation of effects ((Figure 5.3) *income, hispanic, size, etouch, d96pc, b04pc, b00pc*):

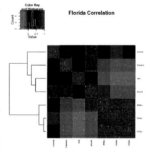

Figure 5.2: correlation of underived variables

---

[3]derived variable

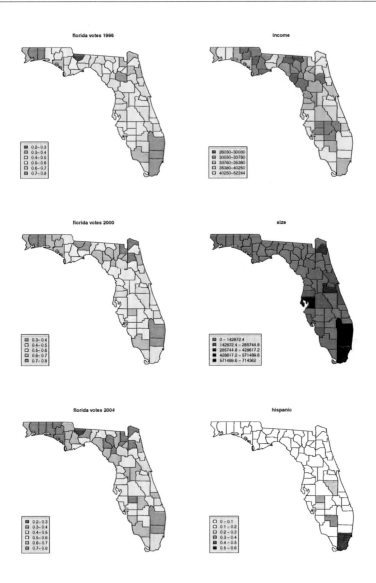

Figure 5.3: Florida elections: geographic distribution of the base parameters

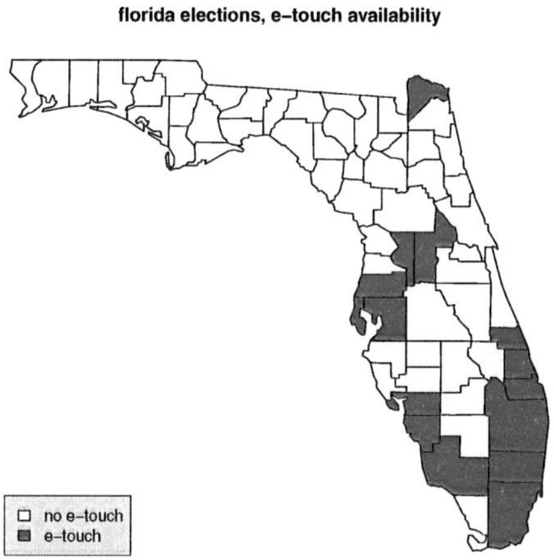

Figure 5.3 shows the data from table 5.1 with respect to the geographic location. The top row displays the temporal voting behavior. Full red colors vote for the republicans where the bright blue color represent democrat votes. The second row shows the areas featuring electronic touch screen devices in green while "classic" voting is shown simply white. Next to *etouch* on the right the number of voters are depicted, the darker the color the more voters were present. In the bottom row the income is illustrated on the left hand side, brighter yellow means higher average beside the percentage of hispanic population on the right.

Looking at the temporal changes in voting behavior in combina-

florida elections, e–touch availability

Figure 5.3: Florida elections: geographic distribution of e-touch availability

tion with the demographic data will presumably not lead to the conclusion that electronic touch screen has an impact on voting behavior. Nevertheless claims exist that there is an effect that skewed the voting in favor of the republican party (Hout et al., 2004). See figure 5.4 on the next page to find out why such a claim could be made. Focussing on the difference between the *etouch* model and the *no − etouch* and *both* might lead to the conclusion that *etouch* linear models differ from the *non − etouch* or *combined* class. An F-test for the difference of variances does not show significance:

```
> modelN=lm(b_change[etouch==0]~d_change[etouch==0])
> modelE=lm(b_change[etouch==1]~d_change[etouch==1])
> modelB=lm(b_change~d_change)
> varN=deviance(modelN)/(dfN<-modelN$df.resid)
> varE=deviance(modelE)/(dfE<-modelE$df.resid)
> varB=deviance(modelB)/(dfB<-modelB$df.resid)
> c("E-Touch"=varE,"no E-Touch"=varN, F=(f<-varE/varN), "Critical
+ value" =  2* if (f<1) pf(f,varE,varN) else pf(1/f,varN,varE))
      E-Touch       no E-Touch          F Critical value
   0.0003935028    0.0006320773    0.6225548317   1.2323942143
```

The F value is smaller than the critical value so this test does not give evidence to use separate models depending on the existence of e-touch.

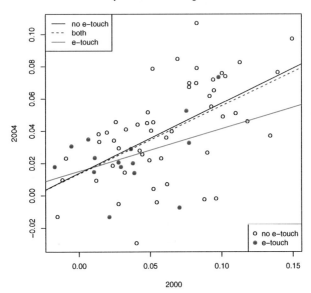

Figure 5.4: comparison of vote changes for the year 2000 and 2004 with respect to e-touch

The above variables will be used in models, still the scaling and influence of the remaining variables is not clear from the geographic plot (figure 5.3). To check if the variables can be used originally for linear modeling take a look at (figs. 5.4).

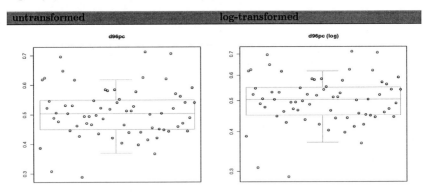

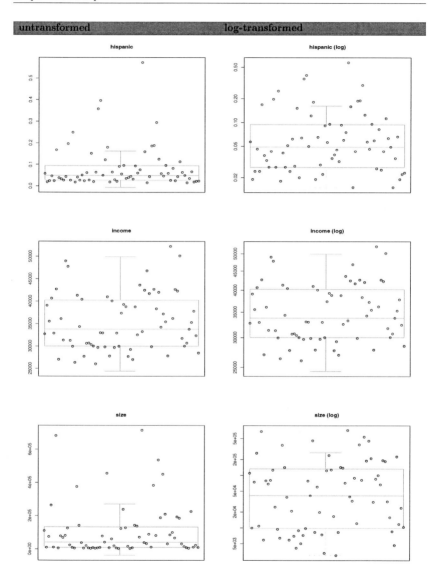

Figure 5.4: Florida elections data set: distribution of the model variables - plain versus log-transformed

(Figs. 5.4) show a heavily skewed distribution for *hispanic* and *size*, *income* is slightly skewed. Linear models can be created from combining log(*hispanic*),log(*size*), either *income* or log(*income*) and *d96pc*. Since the main issue is if there is a notable effect from electronic voting machines, another plot captures the demographic influence data in combination: figure 5.5 (p. 58). The bubble plot shows a combination of *hispanic* versus non-hispanic, who used "etouch" (electronic touch screen) for voting on the x-axis. These y-axis represents the average reported *income* and b_{change} provides the weights. No high income area with above average proportion of hispanics is given, so the data is unbalanced in that way.

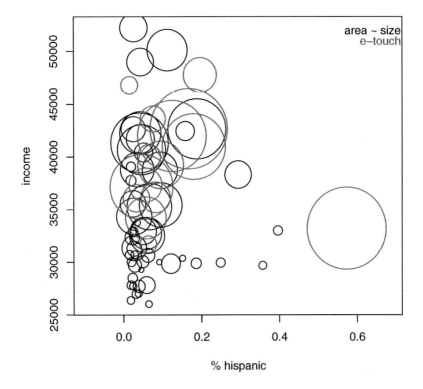

Figure 5.5: demographic distribution of income vs. % hispanic among votes 2004

The relatively large number (as presented by the bottom-right area) of hispanics with low in-come could suggest that in an area with mainly hispanic population huge effort has been taken to provide touch screen devices, but this could also be mere coincidence. Nonetheless, this point needs special care as it is a possible source of bias in the later computed models. Figure 5.5 cannot identify a group that dominates the whole voting. As mentioned above the large mass of hispanics with low income introduces a bias in any models to be discussed. Data dependent bias is important since all calculated models will be skewed. Fitting models from a wrong distribution family supports weak inference at best. Using log-transformation as shown in figure 5.4 (p. 57) seems feasible for linear models - which assumes normal distribution. Electronic touch-screen devices were introduced only over the last years but alas there is no data available about the time when a state received the new devices.

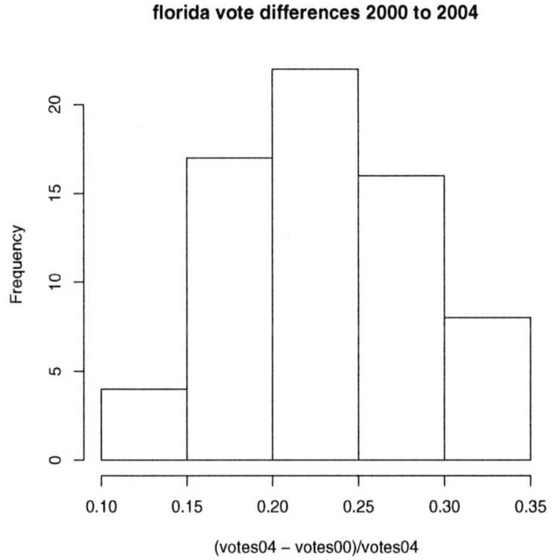

florida vote differences 2000 to 2004

Figure 5.6: distribution of % vote gains from 2000 to 2004

Figure 5.6 displays the average gain of votes from the year 2000 to 2004 as a histogram plot. The average of the additional votes
$mean\left(\frac{\text{votes}_{2004} - \text{votes}_{2000}}{\text{votes}_{2004}}\right)$ is about 23%. This increase is comparable to an additional 19% 1996.

```
> mean((votes04-votes00)/votes04)
[1] 0.2300451
> mean((votes00-votes96)/votes00)
[1] 0.1907217
```

Immigration and other sources of increase in population are the most probable reason for that effect. The correlation of votes over the years stays stable at nearly 100%:

```
> cor(votes96,votes00)
[1] 0.9987252
> cor(votes00,votes04)
[1] 0.9968402
```

Since most of the other variables are derived only one more graphic is shown (figure 5.7). The top part shows for florida a tendency to favor the republican party more strongly over the last three election polls.

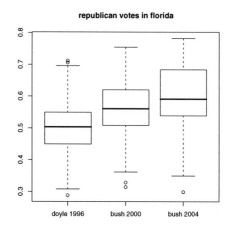

Figure 5.7: votes for presidential elections by year (% republican)

5.2.2 Models

The data has been presented in the last section. Now some models will be fitted using the independent combinations of the given variables *d96pc*, *etouch*, log *income*, log *hispanic* and log *size*. The bias introduced by the southern counties is not found in the linear models, that exhibit the dominating factors. Using standard R backward model selection starting at the full model the selected best model is far too complex:

```
> lm.f104.full<-lm(b_change~d96pc*etouch*log(income)*log(hispanic)
+ *log(size))
> stepAIC(lm.f104.full,direction="both",trace=FALSE)->lm.f104.best
> summary(lm.f104.best)

Call:
lm(formula = b_change ~ etouch:d96pc + log(hispanic) +
    etouch:log(hispanic) + log(income):log(hispanic) +
```

```
    log(income):d96pc + log(size):d96pc + log(size))

Residuals:
      Min         1Q      Median         3Q        Max
-0.0513947 -0.0134510 -0.0004451  0.0132688  0.0418770

Coefficients:
                           Estimate Std. Error t value Pr(>|t|)
(Intercept)               -0.624769   0.142218  -4.393 4.72e-05 ***
log(hispanic)             -0.484156   0.089924  -5.384 1.32e-06 ***
log(size)                  0.053755   0.012598   4.267 7.28e-05 ***
etouch:d96pc              -0.109404   0.042071  -2.600   0.0117 *
etouch:log(hispanic)      -0.019214   0.007826  -2.455   0.0170 *
log(hispanic):log(income)  0.046121   0.008644   5.336 1.59e-06 ***
d96pc:log(income)          0.134307   0.027172   4.943 6.71e-06 ***
d96pc:log(size)           -0.116543   0.025565  -4.559 2.65e-05 ***
---
Signif. codes:  0 *** 0.001 ** 0.01 * 0.05 . 0.1   1

Residual standard error: 0.02043 on 59 degrees of freedom
Multiple R-squared: 0.5543,     Adjusted R-squared: 0.5014
F-statistic: 10.48 on 7 and 59 DF,  p-value: 1.78e-08

> AIC(lm.f104.best)
[1] -321.7194
Residual standard error: 0.02079 on 48 degrees of freedom
Multiple R-squared: 0.6247,Adjusted R-squared: 0.484
F-statistic: 4.439 on 18 and 48 DF,  p-value: 1.826e-05
```

The above variables cover a large search space, considering that there are 5 variables plus 10 interactions. Deeper interactions can be considered in a second step if no good enough model is found. Exhaustive search is not really a good option. The extensive model space is hard to deal with in practice.

With the current tools there is much manual work to be done. Especially handling the amount of models, filtering out bad ones and so on. A practical solution how to constrain model sets is given later in §6.3.2 on page 99.

5.2.3 Intermediate summary

Creating the first models and analyzing summaries are typical steps and show what kind of data is used later on.

1. the data set

2. each command used to analyze the data set

3. analytical plots

4. models

4.1. transformation information

4.2. sampled part of data set

4.3. low level summary data

4.4. intermediate summary data

4.5. residuals

Using this knowledge for the task at hand, to create new software this means:

① everything should be reproducible.

② there should be a way to access the statistics - as mentioned above - for each model to provide a way for model comparison and filtering.

③ the data managed by this software should be accessible for other software too.

④ additional information created for the data set or a model should be manageable at the related object.

⑤ all of the above mentioned data must be manageable with special regard to multi-model inference.

5.3 From Requirements To Design

The preceding example illustrates how much information is contained in a few models. To manage this information by hand is a tedious task. A task to be superseded by a software to manage this amount of model data. From the functions described in the last section the design of the software MORET can be deduced:

5.3.1 Reproducibility

When models are fitted and the results published for papers or other public means the reader expects to recreate exactly the same models as the author originally did. Typing in the identical commands while fitting the original data set does unfortunately not guarantee identical results. There is more than one reason why the results are not identical:

- The software used by the author is no longer available or

- dependent software, e.g. an auxiliary algorithm, changed

- the algorithm utilizes randomization. An example is the classification and regression tree RPART. If the result should be reproducible first *set.seed(..)* must be called before the model is calculated.

Some projects are entirely devoted just to produce reproducible results and provide a way to share these results, see Peng (2008) for instance. MORET is not specialized to distribute any form of analysis but the results - that means models - over the web. Models created with MORET are guaranteed to be stored inside MORETs database in a way that every model can be accessed at this identical *snapshot* level. MORET deals with the reproducibility problem by storing a binary copy of each model into the database, thus allowing another instance of R to reload exactly this model instead of forcing the recreation by script. If the same model command is called a second time MORET will know that this model has been created earlier and instead of re-calculating the same model again the instance created earlier is reloaded. This model database can be shared with other developers so the same results can be guaranteed.

Another crucial part to reproduce a specific model is the way how this model is created. R is not a graphical environment where sliders or drag and drop facilities provide means for model fitting (like e.g DataDesk(Data Description, 1986)). Textual commands define the model, accordingly each command is of importance. MORET keeps track of each command sent to R, so the same model can be fit identically again. The above snapshot method ensures that the model is frozen in an immutable and transferable form. The series of commands, or *script* will recreate an identical model if the same environment is used and randomization parameters are fixed in this script. That way the creation process is supervised. Furthermore the reproducibility of model data is secured twofold.

5.3.2 Access To Model Statistics

A principal advantage of using a database to store models and model statistics is the power of database queries. Database queries allow access to any stored information in a quick way. Additionally the stored information can be accessed in a table-like structure that is well suited for standard statistical software. The challenging part of the database design was how to cope with the different levels of detail as mentioned above. The diversity of available model types posed the second challenge. The first working prototype of MORET supported linear models for a fixed version of R. There are many technical issues that just came into awareness in course of the development. These technical details will be covered later. Right now the focus is the structure of the model: Figure 5.8 reflects access to the stored parts of the extracted model data. As mentioned earlier the model is saved in binary form and additionally the data set used for model creation is stored binary. This UML class diagram shows the different parts of information that are stored in the database.

Each model is built on a special data set. This data set is referenced by the *script*, the series of commands that led to the creation of the model. From the model a global quality statistic as well as a complexity measure is extracted and stored in a database table for easy access and comparability. How this extraction is done technically will be revealed later in the appendix (§B).

The next level of interest is the coefficients in the special case of a linear model. Not every model type provides coefficients. Rather another specific structure like the *cp − table* for RPART models is encountered in most types. Thus each model features a set of *specific values* that can be distinguished by the name.

The most detailed level, the residuals are stored in another table. Not every model will be analyzed in full detail thus the residuals are calculated and stored only when required.

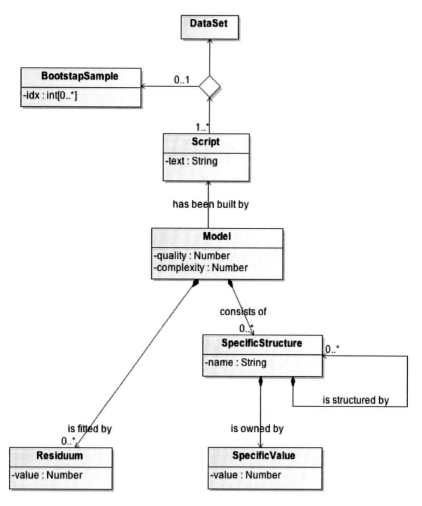

Figure 5.8: UML: overview of classes required to store all model data

5.3.3 Easy Data Accessibility

One requirement is to store every interesting bit of information. Another that this information can be accessed effectively. So any part of the stored data must be accessible in a user friendly way. Figure 5.8 shows most of the information related to a model. Each model - meaning the information in an R model object - consists of high level global statistics and lots of detailed information that comes later. The tasks a user is interested in is to view sets of models, filter these sets according to global or special information or validate models on the lowest possible level. The lowest level is of course the forecast or residual level. Some details about the way a model is fit should be considered, too. A possible way is to compute the model directly from the unmodified data set. Other possibilities involve selecting a predefined subset of the original data, e.g for validation purposes. Lots of special subsets are generated using the bootstrap (Efron and Tibshirani, 1995), (Efron and Tibshirani, 1994). Employing such methods as sub-sampling requires additional information about the data sample. This data sample is not identical to the original data set but a selection of observations. This selection can be stored as an index list where the index represents the original observation.

There are two subtasks for our software to be mastered here:

① Find models by global statistics, filter models by global or special information values

② Store and retrieve the detailed information on a single-model level

5.3.4 Data Accessibility For Other Software

In §5.3.2 (p.63) the internal form of the data has been described. Of course MORET provides access for the data kept inside its database. MORET delegates the model fitting, numeric calculation and form of the structure, to the software R. Unfortunately R does not supply means to re-fetch the extracted statistics from the database[4]. The extracted statistics could be dug out again from the binary models, but that is a laborious task. Especially since all information is stored in the database now.

The problem is to transfer data from the internal database to other software. Even R, which is used by MORET, is an independent module and cannot access the extracted knowledge, located in MORETs database, straight away. To cope with this problem MORET is able to externalize the extracted statistics via files or even drop-and-drag to compatible applications. Most statistical packages support the so-called CSV (comma separated value) format. For other application a structured XML export of all stored data is supported. XSL-T techniques are capable to transfer this format into an arbitrary target format.

Examples will be shown later in the chapters about the solution MORET.

5.3.5 Additional Data

Each type of related data is used most effective with a specialized program. E.g. for a publication usually one program is used to create high quality graphics, while another program like LATEX or a WYSIWYG kind of text compilation program is utilized to write descriptions.

All externally created data relates to one data set, or more specific to one model. MORET stores

---

[4]Of course R offers packages that can extract data from the database again, but the core package does not provide such functionality. Nonetheless the user is forced to learn SQL to extract the knowledge in a usable form.

models and data sets in binary form. There is no technical obstacle that prevents MORET to store any kind of data objects that relates to a data set or model as well. The recent version of MORET manages related data, too. Another R project that addresses a similar problems is *relax*(Wolf, 2008). Relax is more concerned with simple LaTeX document support integration. MORET just stores binary files in its database like the user would do in his file system. In contrast to a file system, which does not hold any information about relations between images, R-code, published documents or working paper all of these are related to a data set or model. MORET manages this logical relation and provides easy access. External specialist software is supported and changes are stored automatically. Without MORET the author is forced to find a manual way to organize all related data efficiently.

5.3.6 Data Organization

Each piece of information that relates to models or data directly has been discussed on the preceding pages.

Basic data manipulation like viewing, analyzing and deleting models or data sets is an integral part of any organizing software and will not be discussed in depth here.

The most valuable aid for an ensemble analyst is the administration of candidate sets. To identify a candidate set a unique group, identified by a unique name, is created. Any model may be assigned to an arbitrary number of groups, either manually or using technical means. These groups play the role of a container and are considered to serve as candidate set. Effective means to select models for groups is a prerequisite.

Apart from candidate sets these groups are useful for other purposes, too. Some models may be subject to a special transformation. Note that model statistics can usually only be compared if the models have been computed on exactly the same data basis. If prior analysis did not reveal what transformation of the data makes the most sense, groups for distinct derived data sets can be utilized. For clearness the analyst is strictly required to rename transformed data! If this is not done diligently models that are not comparable on a theoretical basis end up as candidate set. Due to same parameter names these mixtures are hard to disentangle. The problem at hand is that both models relate to the same set of data. But any model subject to variable transformation has been fit to altered observations and parameter estimates differ. Combining models, that are fit from distinctly transformed data sets, results in strange parameter estimates. To avoid such conflicts the software will not support exactly the same model commands twice. This way data sets and parameters remain unique. More reference to the integrity of the data will be given in a later section.

5.3.7 Summary

Many different tasks have been identified, where a software can actively facilitate the work of an analyst. These range from basic operations like data set and model administration to more comfortable utilities like managing related documents.

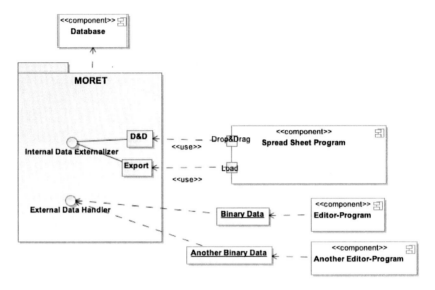

Figure 5.9: UML: MORET and other software as a component diagram

Figure 5.9 depicts the role of the MORET package working with other software tools. MORET uses a database to store external data and keep this data up to date. Additionally MORET provides support for external programs like a spread sheet program. Copy and paste content, drag and drop is usable. Alternatively data may be exported and processed with another tool.

5.3.7.1 Feature Overview

Now is a good point to summarize all the points mentioned above into a short list of features:

① data storage

   ❶ data sets

   ❷ global model statistics

   ❸ special model statistics

   ❹ residuals (low level model statistics)

   ❺ extensions like sub-samples (e.g. from bootstrapping) or transformations

② access stored data

   ❶ search/find model(s)

❷ show overview on global statistics

❸ views on special statistics/interactive comparison

❹ graphics for low level comparison

❺ interface to process raw data with sophisticated tools

③ manage data

❶ delete redundant models

❷ delete data sets

❸ manage names

❹ manage structure (like model grouping)

❺ export data for further processing

5.4 The Development Of MORET

The software specification for MORET has been described in the earlier part of this chapter. Not all of these features have been planned from the beginning. The development was a continuous process that evolved from the first working version:

5.4.1 The Beginning: Trial And Error

At the beginning no additional structures for model grouping, sub-sampling, attachments were foreseen. The application was designed to store exclusively a set of predefined model types. The first prototype featured a database to manage global statistics for any model type and special statistics for linear models. These special statistics cover coefficients and global statistics like R^2 or R^2_{adj}. Other statistics like AIC were added later. Nonetheless a running proof of concept type application was born.

Extracting selected statistics and information from R proved technically challenging. The current version of the software R was 1.8 at that time. Few projects supporting a database combined with a graphical user interface were available at that time. The first idea was to use an interface to communicate with the R software. This interface transfers the data to another programming language where a graphical user interface and a databases is created. Rserve (Urbanek, 2003) was capable to transfer commands to R and return the data to Java - the chosen programming language. Among many other advantages Java works on different operating platforms. Database integration is supported for lots of database systems and many standard GUI components are available. It is not surprising that many statistical tools today are written in Java. For instance http://www.rosuda.org/software contains a list of tools written in Java.

The first issue was to find out how to access the information as mentioned above. Since this chapter is not supposed to reach too deeply into the technical details it is sufficient to mention that the first program was too inflexible. Each value was extracted individually. Moreover the value extraction was defined inside the program code. This may be called *static attribute mapping*.

Extracting single value after single value proved to be too slow for complex models. The first improvement was to use R data structures such as vectors to reduce the overhead calls through Rserve. Each R-model value was statically mapped to a data object. This is the object that is finally stored in the database. After linear models *lm* were successfully implemented, new model types were added. Namely *glm*(generalized linear model) as well as *gam*(generalized additive models) were created. These models feature a common structure: the coefficients. So the effort to create new database tables was related to new specialized tables for unique model type specific attributes:

type	special attributes
lm	R^2, R^2_{adj} (later AIC, $deviance$)
glm	$nulldf$, $deviance$, $nulldeviance$, $dispersion$, aic, $family$, $link$
gam	$nulldf$, $residualdf$, $deviance$, $nulldeviance$, $deviance_{explained}$, R^2_{adj}, gcv, $scale$, n, $family$, $link$

Table 5.2: Model types and attributes

To validate if models with other structure could be processed *rpart* (Classification And Regression Tree (Breiman et al., 1984)) type models were added. This type required not only one new special table since the structure is entirely different but three new tables were added to cope with the tree structure and the $cp - table$.

At this point one flaw of the current design became clear. For each new type of model at least one new database table must be created. If new model types emerge or even worse model attributes change the tables and the values must be redesigned which implies extra effort.

Another point that nearly ended this project was an update of R to the new major version 2.0. Exactly the situation mentioned earlier as extra effort was encountered. Some attributes in R's *gam* were changed and since the mapping was statically coded *gam* could not be processed any more.

5.4.2 The First Working Version

These circumstances finally revealed that the value extraction could not be statically coded. That means that the program part that is responsible for the value extraction process must be separated from the remaining program code. Moreover the specific database structure is subject to redesign into a more flexible structure. As a consequence programmatic adjustments are no longer required. Unfortunately the first working version utilized the sub-optimal special table per model kind approach.

Another aspect of the problem is identified by the absence of a fixed structural description for R model objects. The target structure is defined exactly, but since the source structure is not fixed there is a mismatch in between those objects. To cope with these mismatches the mapping cannot be statically fixed from source to target but another neutral layer must be put in between.

Figure 5.10: Static Direct Mapping

Figure 5.10 shows a direct mapping from a_x to a_x^*. The x indicates the index of the attribute to be mapped and the $*$ denotes a fixed target. It is quite obvious if the n-th attribute is missing something bad will happen. The severity of the problem relates to the kind of attribute and the design. If a mandatory NOT-NULL attribute is missing not a single value can be saved correctly. If this error happens in optional attributes there is non-informative content in the database which is still annoying. Since values are not transformed or observed before they can be written to the database loss of information occurs. If no solution was available MORET would have been cancelled.

An example:
In a tested version the degrees of freedom are hard-coded in the software so a call `summary(#model#)$df` extracts the degrees of freedom. If now for some unknown reason the degrees of freedom attribute in the summary report is renamed to *df.null* the value extraction by `summary(#model#)$df` will not simply return an invalid number but NULL. Worse if this happens in a vector structure where all attributes are considered to be found at a specific index. The whole array will be too short on account of that missing attribute.

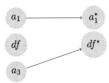

Figure 5.11: Static Direct Mapping: Missing Value

In this example the whole structure is corrupted. a_1 is mapped correctly since it was the first by accident and is still working. For *df* we did get a_3 where we cannot be sure if it is not a string or other incompatible data type. a_3^* will be missing in any case since there is no value left to be mapped from. So there is a maximum of one correct value in this example.

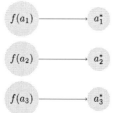

Figure 5.12: Static Configurable Mapping

Figure (5.12) shows the solution for this dilemma. Instead of expecting static values hidden in hard-coded software the value is extracted via a wrapping function that ensures that at least a value of the expected type will be returned. Moreover this wrapping function will not rely on prior and probably out-dated code but a configurable mapping file will define how to fetch that attribute.

In practice there is an additional type validation in the code and to avoid programatic changes in course of a new software version the value-extraction code is placed into a configuration file. For instance a mapping `lm.df={modelname}$df` is defined now. Whenever this attribute changes only one place must be altered - that is this mapping file.

Unfortunately this cannot magically avoid a missing value if the R-object has been changed. But this type of failure is now supervised inside the code. So **if** a value is renamed the new mapping can be adjusted in a text base file without compiling or modifying a single line of code. If there is no replacement value though a dummy value must be mapped and the user must be aware of the fact that this information is no longer available.

5.4.2.1 Analyzing Models

Since MORET can reproduce information even with respect to modification of R-objects, the next interesting challenge is to find models for either a candidate set or one best model for inference. On the global level models will be compared by two global measures, one for complexity and another for quality. This will provide enough information to chose from similar models which models should be considered for further research. The number of models to chose from is very large. If for instance 20000 models have been created by an automatic procedure like combinatoric variable selection, this means a long list to chose from. Assuming that an analyst will work with different types of models as well as different data sets it is a good idea to provide a pre-selection process. Thus the analyst can concentrate on one task at a time. The first version of MORET allowed a selection of data set, target variable and model types as filter criteria to reduce the whole database content to a manageable set. To leverage focus for quality or complexity, filter and sort facilities are provided in this table similar to a spread sheet program. The resulting table (model class, model name, model formula, complexity, quality ..) can be sorted on a single column and each column is filtered using regular expression. It is beyond the scope of this work to familiarize a user regular expressions. In short regular expression are used to validate strings(expressions) using complex but more powerful means as simple substring or

case(in)sensitive rules. Regular expressions relate to automata theory and can be mapped to a
state graph. Like this graph the expression is matched against a state or in case of the expression
a string. See for instance McNaughton and Yamada (1960).

Regular expressions can produce a filter based on a match of string of characters. Since model
formula, model name can be written down as characters an arbitrary expression can be generated
for either a special model parameter or anything that is written down as a string. Please refer
to the appendix (§B.3) for an introduction how to take advantage of regular expressions.

Model Table - Global Selection   Using the means described above, a global model table like
5.3 is where to start selecting models for candidate sets:

Model	R Command	Quality[1]	Complexity[2]
lm.1	lm(b_change~d96pc+hispanic+income+etouch)	0.0374	62.0
lm.2	lm(b_change~d96pc*hispanic*income*etouch)	0.0291	51.0
glm.1	glm(b_change~d96pc+hispanic+income+etouch)	0.0374	62.0
⋮	⋮	⋮	⋮

Table 5.3: Exemplary global overview table, displays high level summary attributes of a model
like quality and complexity

In practice this table will be much longer but since these models have been introduced earlier
in §5.2.2 this provides a concise example. The first point that catches the eye is that another
quality statistic than R^2 is used. Instead *deviance* serves as global quality measure. For linear
models this measure is rarely used, but deviance is available for linear models and generalized
linear models as well. This permits comparison on a common quality criterion. Usually there is
more than one way to select a global quality statistic. Another possibility is to use *AIC*, which
combines complexity and quality.

The next step towards candidate set selection is to compare models at various benchmarks. At
this level of detail other global statistics may be compared, or at greater detail the coefficients.
R summary reports provide the desired information in a textual way. Nevertheless it is a tedious
task to compare more than two to three models solely based on verbose text summaries.

Model Tree - Detail Inspection   An alternative way of displaying information is to present the
content as a tree. Using computers everybody has seen file-structure trees. R's standard textual
summary report looks like the following

```
lm(formula = b_change ~ d96pc + hispanic + income + etouch)

Residuals:
      Min        1Q    Median        3Q       Max
-0.058816 -0.013900 -0.001767  0.015640  0.059912
```

---

[1]deviance

[2]df

```
Coefficients:
             Estimate Std. Error t value Pr(>|t|)
(Intercept) 7.153e-02  2.207e-02   3.241  0.00192 **
d96pc       1.114e-01  4.064e-02   2.740  0.00801 **
hispanic   -5.095e-02  3.155e-02  -1.615  0.11145
income     -2.452e-06  5.603e-07  -4.376 4.73e-05 ***
etouch      5.250e-04  7.976e-03   0.066  0.94773
---
Signif. codes:  0 '***' 0.001 '**' 0.01 '*' 0.05 '.' 0.1 ' ' 1

Residual standard error: 0.02455 on 62 degrees of freedom
Multiple R-Squared: 0.3239,      Adjusted R-squared: 0.2803
F-statistic: 7.426 on 4 and 62 DF,  p-value: 5.929e-05
```

This presentation shows all information that the programmer deemed important for an efficient overview. At the bottom is a text block regarding global statistics. Above a block regarding the intermediate information level, the coefficients. On top of the coefficients some information concerning the distribution of the residuals. First of all the model formula is displayed for reference. Each part that is presented is useful. The order and importance can be argued about, but each user soon gets used to this kind of display. Let us transform this information to a *tree representation*, regarding the structure as shown in figure 5.8. At a first glance figure 5.13 contains exactly the same summary information transformed into another graphical representation. In fact the plain textual information has been enriched by the new structure. The level, that is the number of edges on the tree-graph, reflects global importance. Another minor importance criterion is the order of node-siblings. Primarily the sibling relation expresses that information on this level belongs to one object. Using the concept of the *tree-path* any node can be identified by its parent and the child number.

That provides means to identify comparable values on different models. The path */S/Coefficients/income/probability* for instance identifies a model specific value. In the model displayed in figure 5.14 that value is $4.73e - 05$. The same path for *lm.2* (cf. table 5.3) - that is the full model - yields 0.578.

After a plain text summary has been transformed into a structural representation an abstract representation for common information is now available. Of course this information can be found in R, too:

```
> summary(lm.2)$coefficients[which(rownames(summary(lm.2)$coefficients)
+ =="income"), which(colnames(summary(lm.2)$coefficients)=="Pr(>|t|)")]
[1] 0.5782663
```

This command is hardly readable. Breaking it down into bits unravels what happens:

```
> #create alias name for named coefficient matrix
> lm2coeffs<-summary(lm.2)$coefficients
> #find out which row contains the coefficient named "income"
> incomeIdx<-which(rownames(lm2coeffs)=="income")
> #find out which column contains the coefficient p-value
```

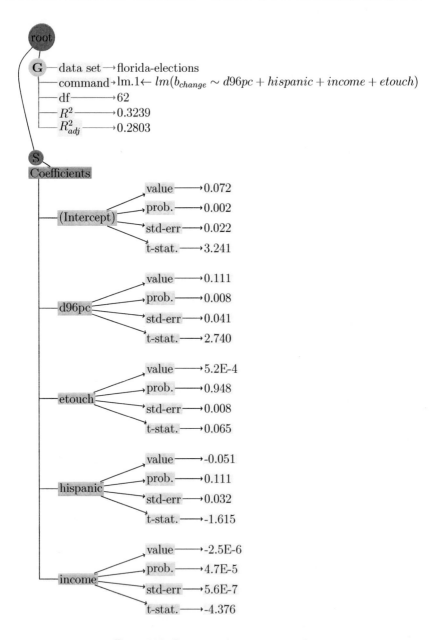

Figure 5.13: Summary report as tree mapping

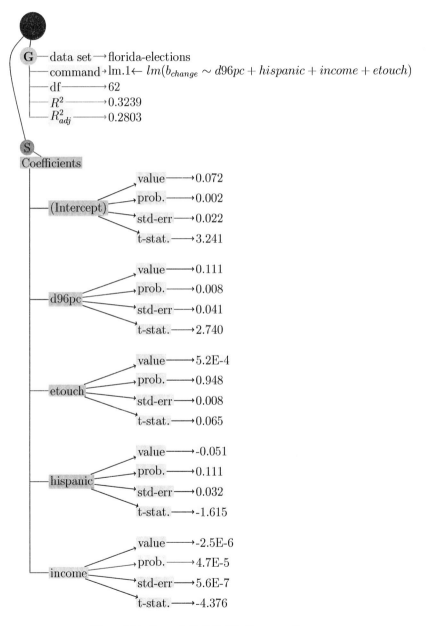

Figure 5.14: Example Path In The Summary Tree

```
> probabilityIdx<-which(colnames(lm2coeffs)=="Pr(>|t|)")
> lm2coeffs[incomeIdx,probabilityIdx]
[1] 0.5782663
```

Conceptually the information can be found in R in an adequate way, but the syntax, calling each node by name is more convenient.

Multi-Tree And Model Comparison

This structure supports interactive model comparison on the global and intermediate detail level. Theoretically residuals or fitted values can be compared, using the next algorithms. But the selection of so many nodes is inconvenient.

The first algorithm deals with the creation of a *selection multi tree* from a set of structure trees. Given models in tree representation like figure 5.13 some models share at least part of their structure. The goal is to compare models on as many benchmarks as possible. The intersection of common benchmarks is the foundation to compare models. Tree like models and linear models for one instance do not share many properties. Coefficient models (lm, glm, gam) may be compared on coefficients and common global statistics. Finally residuals can be compared for any model type.

The final algorithm 2 (on page 78) compares node path values, so either the model features that property or it does not. These node paths may be selected interactively as well as the set of models. In between algorithm 1 transforms a set of models into the set of all node paths: the selection multi tree. This multi tree serves as an interactive selector for the properties to be compared.

Algorithm 1: Create Selection Multi Tree

Input: $\mathcal{T}$ a set of model trees
Output: $\mathcal{R}$ the multi tree
initialize $\mathcal{R} := root$;
foreach $\tau \in \mathcal{T}$ **do**
  | Copy each structure node n recursively starting at $\mathcal{R}$;
end

The result of algorithm 1 produces a SMT as shown in figure 5.15. While this figure is crowded it displays every distinct attribute one model has. Since $lm(b_{change} \sim d96pc * hispanic * income * etouch)$ even covers all attributes including interaction terms the selection tree for this model is equally complex. Now the good news is that this complex model contains all sub-models. Since many attributes are shared it is possible to select a set of tree-paths from the SMT and compare a set of models on this selection.

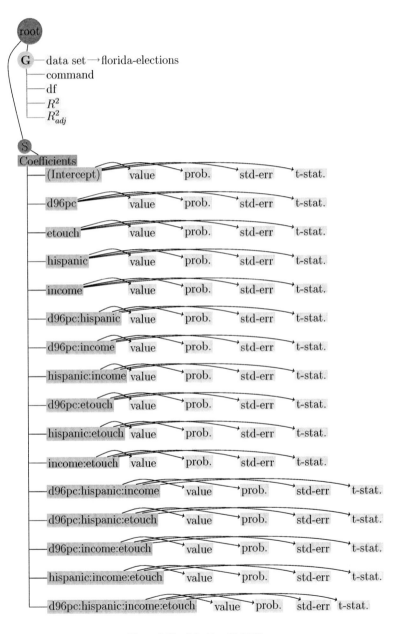

Figure 5.15: Selection Multi Tree

Algorithm 2: Multi Tree To Table Transformation

Input: $\mathcal{T}$ a set of model trees;
$\mathcal{P}$ a set of selected tree paths from SMT
Output: table containing tree data
initialize $column := 1$;
initialize $row := 1$;
initialize $maxcolumns := length(\mathcal{P})$;
initialize table header $\mathcal{H}$ as follows:
foreach $\rho \in \mathcal{P}$ **do**
$\quad$ | $\mathcal{H}[column] := toString(\rho)$;
$\quad$ | column:=column+1;
end
foreach $\tau \in \mathcal{T}$ **do**
$\quad$ | **foreach** $\rho \in \mathcal{P}$ **do**
$\quad\quad$ | | **if** τ *contains treepath* ρ **then**
$\quad\quad\quad$ | | | set value at table[row,colum]:=value($\tau[\rho]$)
$\quad\quad$ | | **end**
$\quad\quad$ | | ;
$\quad$ | **end**
$\quad$ | $row := row + 1$;
end

Using a selection of all terms without interactions, and focussing on the value and p value algorithm 2 will produce the following:

Input:
$\mathcal{T} := \{lm(b_{change} \sim d96pc + hispanic + income + etouch, lm(b_{change} \sim d96pc * hispanic * income * etouch\}$
$\mathcal{P} := \{$ S/Coefficients/(Intercept)/value, S/Coefficients/(Intercept)/probability,
S/Coefficients/d96pc/value, S/Coefficients/d96pc/probability,
S/Coefficients/etouch/value, S/Coefficients/etouch/probability,
S/Coefficients/hispanic/value, S/Coefficients/hispanc/probability,
S/Coefficients/income/value, S/Coefficients/income/probability $\}$

Output:
Querying those models for other coefficients yields an empty cell for $lm(b_{change} \sim d96pc + hispanic + income + etouch)$ but a value from $lm(b_{change} \sim d96pc * hispanic * income * etouch)$. Manually it is quite a laborious to extract, sort and insert these values into a table (cf. table 5.4). Even for two or three or even a small set of models manual processing is inconvenient. Software is required to provide the managing part, like extraction, sorting and table creation. Especially interactive selection from a SMT is an helpful asset.

Taking Advantage Of The Selection-Table

The selection-table 5.4 contains the values of all models with respect to the selection of the SMT. This interactive process and a real MORET example will be shown in the next chapter. Other useable tools while working with that table involve interactive plots or more detailed access to summary statistics from a selection. The latter task is either delegated to a spreadsheet processing program or another specialized statistic software. For exhaustive analysis a maximum SMT and a full selection table without manual interference is used. To take advantage of that

| model | $/Coefficients/ | | | |
| | (Intercept) | | d96pc | |
	value	prob.	value	prob.
$lm(b_{change} \sim$ $d96pc + hispa + incom + etouch)$	$7.153e - 02$	0.00192	$1.114e - 01$	0.00801
$lm(b_{change} \sim$ $d96pc * hispa * incom * etouch)$	$6.329e - 02$	0.764	$1.681e - 01$	0.668

| model | $/Coefficients/ | | | | | |
| | etouch | | hispanic | | income | |
	value	prob.	value	prob.	value	prob.
+	$5.250e - 04$	0.94773	$-5.095e - 02$	0.11145	$-2.452e - 06$	$4.73e - 05$
*	$1.272e + 00$	0.149	$-2.686e + 00$	0.377	$-3.312e - 06$	0.578

Table 5.4: Manual example for a value-table, this supports comparison of different models and types as well

information all stored models information must be transformed in a format compatible to the analysis software; so called CSV (comma separated value) is supported by a wide range of software.

Summary For The First Version

The first version of MORET is able to fulfill many tedious tasks, when done by hand. Looking back at the requirements section §5.3 there are still some issues to be solved. For example generic mapping for other model types, extended information like bootstrapped sub-samples, more sophisticated possibilities to select model-sets remain unresolved. Nevertheless this first version, introduced at useR 2006 in Vienna, Austria was already able to solve many headaches of model set administration.

Further interesting features that were at least at an early stage involve

☞ a graphical model creation wizard to support mass model creation by selecting target variable, model variables and model type.

☞ a batch file process is available to support reproducible results. R commands from a file are processed line after line.

A MySQL database was required and the mappings had been statically adjusted to the current R version.

5.4.3 Advanced Features

The first working MORET version facilitates many tasks that require lots of manual work. Primarily the manual process of collecting special model information is now effectively accomplished using the various overview tables (5.3, 5.4) combined with SMT selection (see algorithm 2). Before the generic database structure and the problems how to configure mappings for this structure can be implemented a technical annoyance must be dealt with first: Any software that relies on a database needs a specialist to care for this database. MySQL is famous for its performance and effortlessness to set up. Nonetheless there is need for an installation and the tool MORET that depends on this installation must be configured properly. The database driver must be compatible with the installed database version and so on. These are technical obstacles that must be dealt with first before the database is working as expected to make MORET work.

5.4.3.1 A Self-Configuring Database

Fortunately for Java some database projects exist that need no special setup and administration. Namely the apache DERBY-DB (Group, 2004) and the HSQLDB project (Group, 1995). The latter one was chosen since the project was smaller and the first experiments looked very promising.

Using a pure java database holds the advantage that the database configuration can be completed without external dependencies. That means neither administration nor a specialist is required. The database is created automatically when MORET starts for the first time. Some details regarding this database involve the setup to run as a server and store values immediately. The online documentation at the HSQLDB project URL (Group, 1995) covers these detail exhaustively.

Thus MORET is enabled to create and configure an internal database, and keep it up to speed. Even extensions for later version are processed automatically - which prove a major problem for other growing projects. When the database and the software are managed independently, two steps are required to synchronize both. In contrast combining both leverage easy upgrades. HSQLDB supports standard SQL like many other databases. So anyone skilled using SQL may retrieve all information directly from this database.

In summary the pure java database offers the following advantages:

✔ independence from external database configuration

✔ standard SQL support

✔ automatic upgrades to new MORET versions supported

✔ easy optimization because the software management is self-contained

Comparison to an external standard database

If a database administrator is available and JDBC drivers are at hand, this special database might be adjusted and kept up to date. The technical obstacles of working with different databases can be overcome. Minor adjustments of database queries must be performed by this specialist. Since the latest release, MORET checks its database consistency at start-up. Most alterations can be finished automatically. The remaining few are guided by messages about the required changes. In most cases an automatic update is possible; nevertheless a specialist is required for the rest. Any advantage should be considered and weighed carefully. Convenience against approved stability is not easy to choose since both have their intrinsic advantages and drawbacks.

5.4.3.2 Generic Database Structure

The required database for MORET can be guaranteed, using a pure java database by default. No specialist is required to install, configure or manage this database. The next challenge is the definition of a generic database structure that is able to store any possible model type.

Figure 5.8 (on page 64) displays a structure for any possible model type. This structure is utilized to generate an overview table 5.3 (on page 72). In the last section a tree presentation for model summaries was proposed 5.13 (on page 74). Any model can be transformed into this tree representation. The tree structure (cf. figure 5.8) is not model specific but universal. The *specific structure* needs closer examination, though.

To keep the tree persistent both, graph structure and values must be stored in the database. The nodes from this graph can be distinguished by name, a neighbor node and a given order. Values are of special different types - *strings, integer* or *float* numbers as well as *boolean* types are found. When the order of these values is fixed, too, any possible data type and structure can be stored. Figure 5.16 shows the structure. The interface *VertexValue* is used to emphasize the different types of nodes.

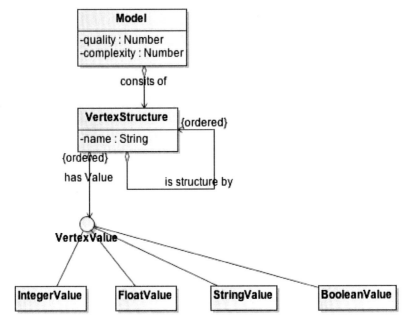

Figure 5.16: UML class diagram: a generic structure that is capable of keeping any structured value

Different types are easily mapped in a programming language like *java*. The database design requires more thought. The values belong to at least one table. Either the referring structure is stored in a big table that supports all types or a special table per type. Remembering that a tree is an acyclic graph this design, as shown in figure 5.16, can be broken down to a *vertex* table that is able to store any value and an *edge*-table to store the structure. The vertex table uses an order attribute, a type attribute and supports one value into a column according to the type attribute.

vertex
- order:int
- type:enum
- stringvalue:string
- intvalue:int
- doublevalue:double

Table 5.5: vertex attributes in UML class notation

Boolean values, that are usually not a database basic type, can either be written into a string field or an integer field. Number conversion - into a string - is possible in practice but it is not a good design because of the type mismatch. The database queries on number-strings yield awkward results.

This design is capable of storing an arbitrary java object. One problem remains: how to get these values from a model object and transform these to the defined structure.

5.4.3.3 Model Configuration

Alas no common, reliable attribute exists to determine what R-object represents a model. The $call object is an indicator but not compulsory. Analyzing objects in the R workspace is also too costly as the model object can be "hidden" inside R *vectors* or *lists*. These structures are generic and may contain any R-object. Furthermore the model objects or summary objects may contain too much information, that will not be needed for model analysis - like a numeric algorithm identifier or parameter used to stop iteration. Usually the model values are the focus of selection strategies, not internal algorithmic criteria. This problem is closely related to *static configuration mapping*, see figure 5.12 (on page 71).

In this context the mapping must be considered in a more general framework. The source is the model summary object, to be more specific the *summary model tree*. This tree representation is accessible from R through Rserve and the complete tree could be stored into the database. In practice some model trees are very complex. Pruning the tree before the tree model is stored is a good idea.

Two requirements are to be implemented here:

① support to configure what a model is and determine the structure and values that must be saved into the generic structure

② support model tree pruning, restructuring, and reordering

Furthermore the whole configuration process should be supported by a graphical user interface. The first step of this configuration process can be done using a tree construction algorithm that defines a tree in a structure which can be stored into the database. Algorithm 3 describes the operation of this algorithm, the transformation of nodes or values is not mysterious but a technical process that converts an R object to the generic structure.

Algorithm 3 transforms the given data $\mathcal{T}$, which is the specific java object given by Rserve into

---

Algorithm 3: Tree To Graph Transformation

Input: $\mathcal{T}$ a model tree in the form of Rserve specific types, $\mathcal{G}(v, e)$ the model graph. initially $\emptyset$, ρ the parent node, initially *root*

;

Output: $\mathcal{G}(v, e)$

foreach $\nu \in \mathcal{T}$ **do**

    if ν *is leaf* **then**

        $v^* := transfromToValue(\nu)$;

        $e^* := (\rho, v^*)$;

        $\mathcal{G}(v, e) \uplus (v^*, e^*)$;

    else

        $\hat{v} := transfromToStructureNode(\nu)$;

        $\hat{e} := (\rho, \hat{v})$;

        $\mathcal{G}(v, e) \uplus (\hat{v}, \hat{e})$;

        recursively call algorithm 3)($\mathcal{T}$,$\mathcal{G}(v, e)$,$\hat{v}$;

    end

end

---

an internal tree representation. Note that the output graph $\mathcal{G}(v, e)$ is a set of distinctive nodes and furthermore we need to ensure that this algorithm 3 produces not only a graph but a tree. Each node must be uniquely identifiable by its *tree-path* as mentioned earlier (ref. figure 5.14 on page 75). Programmatically, e.g. using java *SortedSets* of nodes, the problem is solved easily. Sorting is important for presentation reasons and a minor issue for the data. Uniqueness is really important for clarity and thus later to compare these distinctive models.

For the unique tree/graph $\mathcal{G}(v, e)$ there are further rules to obey after this initial transformation. Initially $\mathcal{G}(v, e)$ contains all information given from the source but in a raw form. The command that created this graph can be saved and used in the same way as the hard coded types *lm*, *glm*, *gam* or *rpart* mentioned earlier. The task still remaining is to restructure this tree. Apart from creating a graphical tool for these structural changes 4 basic operations suffice to restructure the raw tree:

1. remove any node

2. create a new structure node

3. unlink a node from its current position (that is to remove the edge between the node and another one)

4. to link a node to a new associated node (analog to create an edge between those two)

With these operations two other operations come for free:

1. rename any structural node

2. change the order (which is reflected in the sequence of edges)

5.4.4 Miscellaneous Features

More useful functions evolved with the progress or the project. Due to the fact that some features are helpful the following lists a small selection and presentation of further ideas:

Database Export And Import

There are advantages if the database is just located on one computer. For example no network connection is required for database access. But when the database content must be available for more than one researcher this setup is troublesome. There are two possible solutions to cope with that requirement:

① create a shared database instance that can be accessed by everybody over the net at the cost of network dependency

② support data export and import

The first option required network and database specialists.
The second option is interesting since it allows many applications only a specialist could provide otherwise:

export and import for data security To transfer data to another database instance a dump (that means all the data content in a technical format) is created. This dump can be transferred from one database to another instance. This is useful if no shared database is available for all participants. Or another use is to migrate the database from one computer to another one. Importing into an empty database is easy. Sophisticated methods for importing into a non-empty database are available, too.

permission management Since only a part can be imported (technically by using a filtered export) only parts of the data/models need to be exported and thus can be analyzed by the shared parties.

database migration If ever the database needs fundamental changes or there is an administrator available the old data can be simply imported/exported from a neutral format (the dump) into the target database

5.4.4.1 Finding Models By Values

The next concept takes advantage of the fact that every model value can be accessed by its unique *tree-path*. A tree-path has been shown in the example figure 5.14 (on page 75). The model value is found at the leaf of this path, thus

```
Coefficients/income/prob
```

provides the probability of the income coefficient. Using this example tree-path selection on a model can only provide a value if the model has coefficients and one of these coefficients is named *income*.

This idea can be used for another way of choosing model sets. If a model features a coefficient named *income* all of these models can be selected. A means to chose an attribute, the tree path must be provided, additionally - according to the attribute type *string, boolean* or *numeric* the selection is made. Of course attribute combinations are useful for controlled model selection. For the two models mentioned above the following queries can be created:

Query "find models that" ...	Result
features a coefficient named *size*	$\emptyset$
provide coefficients with *probability* $\leq 5E - 5$	$\{d96pc + etouch + hispanic + income\}$
feature a coefficient named *income*	$\{d96pc + etouch + hispanic + income,$ $d96pc * etouch * hispanic * income\}$
has $R^2_{adj} > 0.45$ and features *etouch*	$\{d96pc * etouch * hispanic * income\}$
$\vdots$	$\vdots$

Table 5.6: Feature queries, some examples in verbose common language

This example (table 5.6) is artificially constructed of course. Nonetheless the usefulness of this idea becomes evident. In practice the set of models will never be as small as two and when there are thousands of models using this *feature query* provides a great way to select subsets for candidate set selection. Using a candidate criterion as must-have feature narrows the search space effectively.

Model Groups

Model sets can be found using *feature queries* as mentioned just above or by using manual selection from the model table 5.3 (see page 72). Especially the manual process is time consuming and since the resulting selection will be used as a candidate set this step is crucial. Interactive model selection must be done to select models effectively. This selection can be assigned a common name, used to identify the candidate set.

Alternatively this *model group* identifier can be used to identify special common features among models, like a must-have coefficient. Using groups that way a candidate set is selected by a logical combination of groups. As an example the candidate set is chosen by a combination where a must-have coefficient is present AND no sub-samples have been chosen. Since the selection result can be filtered out manually again this pre-selection process is bound to narrow down to a meaningful candidate set.

R Extensions

A minor issue to solve was to provide a way for non black-box sub-sampling (bootstrapping). The bootstrap selects a subset of the original data but to support accurate analysis of a single bootstrapped model the bootstrapped source data must be available. To do this there is a way to integrate plain R functions into MORET and this function return the index set of the selected observations (rows in the data matrix). This index set can be displayed, for example by analyzing missing observations. Potential outliers can be found from a new perspective. (See figure 4.1 on p. 42 for example). Another advantage is that sealed bootstrap procedures can be analyzed transparently single model by model.

Other potential for R extensions is the integration of other black-box procedures like boosting (see §3.3.3 on page 28). When MORET is control of the boosting process all intermediate models can be stored in the database and analyzed with the provided tools.

Additional Data

All data mentioned earlier can be managed using MORET. Presenting the knowledge won is another issue. Graphics will be created to summarize the data, text notes will be written. Even papers or larger works involving combination of graphic and text can be fed by the knowledge contained in MORET's database.

All of these single pieces are linked to either a model or a data set - since that is the main part of MORET's data. To cope with such material in a way that all knowledge can be kept in one place linking any computer based document to the actual data is supported. These documents will be integrated into the database - taking advantage of a database and the possibilities of import and export. Techniques are available to even keep the stored data in the database at the newest available state while providing accessibility to system editors. In that way MORET can be used like the file system but the folders are the models and data sets.

5.4.5 Summary

Many ideas were presented in this chapter. Beginning with an example data-set to familiarize with the problem of model selection. The administration of these models will become tedious regarding the growing amount of data that has to be handled manually:

- ✔ data set
- ✔ transformations
- ✔ the model and all model values
- ✔ plots used to analyze the data
- ✔ specific additions to models or data sets.

Each of the above mentioned parts can be automatically administrated using MORET.

Every part of the data will be automatically stored in a database where each transformation step will be kept reproducibly and the information can be accessed more easily than from any other form of data storage. By restructuring the model to a tree representation new benefits arise. The model values can be compared using this tree-path selection. Since the data is linked in the database the binary form of the model is coupled with the values - even additional files can be considered special values - and vice versa. Thus the binary model can be used in R as usual with the additional benefits gained by the databases tools.

At the end of this chapter special functions involving better usability are covered. Data retrieval, data safety, data deployment and more were shortly mentioned. Since it is technically demanding to manage a database centrally there are many handy additions for sharing parts of a local database.

Gathering all these advantages in this summary it is clear that MORET offers great improvements handling single models or model ensembles.

Chapter 6

Working with MORET

6.1 Overview

In the preceding chapters concepts for statistical modeling have been presented. A wish list has
been assembled how a software can assist an analyst, with special regard to model ensembles.
This practical chapter presents some technical details regarding the implementation of MORET.
Recommendations are given how to best apply the new tools.

An example is given how to adapt an arbitrary model leveraging the *generic* model type. Candidate set administration is discussed and the features used for this task are explained. Next
is another practical example about candidate set selection. MORET's tools are shown in action and explained. The final section shows MORET in combination with external software.
A spreadsheet program is combined in one example and another shows interactive plots with
Mondrian.

6.1.1 Data Set Handling

The foundation of the software MORET has been presented in the last chapter. Looking at the
start-screen it presents like any ordinary GUI written in java. The major part is occupied by
a feedback window that tells the user what is going on and an input field below. Many menus
facilitate everyday use and lots of specific functions can be found on the top menu bar. Since
MORET has been designed to manage models and ensembles, some very specific restrictions
apply. These restrictions are covered in §6.1.2 (p. 88). A simple example of common steps is
given next: The first thing an analyst will usually start with is to load and analyze a data set.
Choosing a data set can be done using a graphical wizard. This graphical wizard will create the
R command that could otherwise be entered directly:

```
read.table(election.txt,header=TRUE)
```

Loading data via a GUI is nothing extraordinary. MORET performs some extra efforts automatically: A binary copy of the data set is stored in the database. Additionally MORET assumes
that any model created is related to the data set just loaded. The bottom line in the GUI

shows the current data set. This focus is very important, to work effectively with one data set. Technical considerations must be taken into account, too. A second command is automatically produced and executed:

```
attach(election.txt)
```

The `attach`-command is executed automatically focussing on one data set. Using this data set different plots (interactive plots are supported since iPlots (Urbanek and Wichtrey, 2003) collaborates with MORET via Rserve) in the pre-analysis step determine what kind of models are to be considered. In the next step models are fit using known *intercepted* model commands. Either manual input or a step-by-step wizard is used. *Intercepted* in this context means that the command, as typed in by the user, will not be executed in R literally. Internal processes are triggered at the same time - a technical explanation will be given later in §B.1.3 on page 153. No adjustment is needed from the user's point of view. Entering R-commands as usual fits models. Creating *stored* MORET models requires no additional effort. A simple example how this wizard assists the user creating sets of models is shown in §6.3 on page 94.

6.1.2 The Importance Of Data Integrity

Creating a good candidate set from a single data set is a demanding task. That is the reason why MORET permits only a single data set to be processed at one time.

On the one hand this imposes a restriction. Permitting only a single data set for model creation enforces that all results relate to this data set. On the other hand this technical constraint is essential for MORET to link models to this data set. Concurrent sets of data depend on filters that decide for any command which data set is affected. The model generation process needs to be traceable. This is the main reason to impose this constraint. Additionally a filter that is able to correctly relate commands for different data sets for one or more models cannot be created. The series of commands that is required to create one model is referred to as *script*.

A short example sketches how concurrent data sets prevent model - data set relation :

```
election<-read.table(election.txt)
florida<-read.table(florida04.txt)
```

... before models are fitted standard plots will be preformed. Transformations may be used and new derived variables introduced. ...

```
lm.election.1<-lm(d96pc+hispanic+income+etouch,data=election)
lm.florida.1<-lm(d96pc+hispanic+income+etouch,data= florida)
```

Now the *script* for `lm.election.1` contains any command entered after loading the data set. Unless a very sophisticated filtering algorithm can sort out which part of the commands belongs in the *script* that led to a model. Such a *script* separation algorithm will be very hard to implement. The transformation commands may use intermediate objects that can relate to more than a single object at once. Consequently these parts are related to all models. `lm.florida.1` now contains the same *script* as `lm.election.1` when the user input is not be separated. The script is not related correctly to each data set for merely two data sets. More data sets aggravate the situation. As a result each model *script* is long, unreadable and incoherent.

Regarding the model structure as earlier shown in figure 5.8 on page 64 each model consists of a *script* part and the data part. This *script* part, as defined above, represents any command that contributed to the final model. Usually diagnostic plots and transformations are important. These commands contribute valuable information to the created model and must be stored, too. The *script* needs to be able to recreate the same model - disregarding technical problems like random seeds. Next is a short example that demonstrates this technical problem when a *script* - without additional precautions - is not sufficient to reproduce exactly the same model. This problem can be obviated by storing the binary model. Any time when the same model command is executed again, the model stored previously is reloaded from the binary snapshot.

6.1.3 A Textbook Example

Introducing CART (Breiman et al., 1984) the *iris* data set is commonly used. This data set contains measurements in centimeters for the variables sepal length and width and petal length and width, respectively, for 50 flowers. Each flower is from one of 3 species of iris, either "Iris Setosa", "Versicolor" or "Virginica". Venables and Ripley (1998, pages 421+) provides one analysis using R's *tree* algorithm. We will use the *rpart* algorithm for this example. Figure 6.1 shows a partitioning of the space among these iris flowers. Green color dots represent the flowers of species "Setosa", blue is "Versicolor", and red "Virginica". The axis display the dimensions petal length and width.

An identical model is usually created from identical input (the script). In this case the script consists of a one line command. For the *rpart* model the assumption of identical values does not hold true:

```
> library(rpart)
> data(iris)
> iris.rp1<-rpart(Species~.,data=iris)
> iris.rp2<-rpart(Species~.,data=iris)
> printcp(iris.rp1)

Classification tree:
rpart(formula = Species ~ ., data = iris)

Variables actually used in tree construction:
[1] Petal.Length Petal.Width

Root node error: 100/150 = 0.66667

n= 150

    CP nsplit rel error xerror     xstd
1 0.50      0      1.00   1.17 0.050735
2 0.44      1      0.50   0.70 0.061101
3 0.01      2      0.06   0.08 0.027520
```

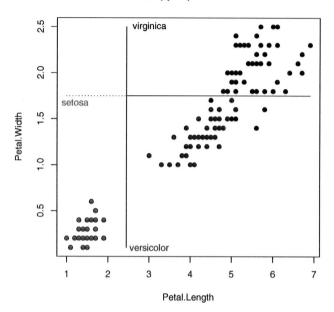

Figure 6.1: CART example: the *iris* data set is partitioned using the *rpart*-algorithm

```
> printcp(iris.rp2)

Classification tree:
rpart(formula = Species ~ ., data = iris)

Variables actually used in tree construction:
[1] Petal.Length Petal.Width

Root node error: 100/150 = 0.66667

n= 150

      CP nsplit rel error xerror     xstd
1 0.50      0      1.00    1.19 0.049592
2 0.44      1      0.50    0.65 0.060690
```

```
3 0.01      2      0.06    0.09 0.029086
```

On closer inspection this example reveals that the $cp-table$ differs in the last two columns. The identical command

```
rpart(Species~.,data=iris)
```

produced two non-identical models. In this case the different result was a direct result of a random generator. The computation of models in general involves numeric algorithms that may have to deal with ill conditioned matrices - so an identical result cannot be guaranteed in all cases. Fitting algorithms become more stable over time but to reproduce identical results on all kind of computers is not very probable. In this simple example we can fix the random number generator and reproduce the calculation:

```
> set.seed(1)
> iris.rp1<-rpart(formula=Species~.,data=iris)
> set.seed(1)
> iris.rp2<-rpart(formula=Species~.,data=iris)
> printcp(iris.rp1)

Classification tree:
rpart(formula = Species ~ ., data = iris)

Variables actually used in tree construction:
[1] Petal.Length Petal.Width

Root node error: 100/150 = 0.66667

n= 150

    CP nsplit rel error xerror     xstd
1 0.50      0     1.00   1.17 0.050735
2 0.44      1     0.50   0.73 0.061215
3 0.01      2     0.06   0.09 0.029086
> printcp(iris.rp2)

Classification tree:
rpart(formula = Species ~ ., data = iris)

Variables actually used in tree construction:
[1] Petal.Length Petal.Width

Root node error: 100/150 = 0.66667

n= 150

    CP nsplit rel error xerror     xstd
```

```
1 0.50      0      1.00    1.17 0.050735
2 0.44      1      0.50    0.73 0.061215
3 0.01      2      0.06    0.09 0.029086
```

Identical values from an arbitrary script cannot be guaranteed. Special care needs to be taken from the user's side. MORET solves with this problem by writing binary model files into the database. If the user tries to fit the identical model again the model is not computed. Rather the binary copy will be reloaded. When the same model command is entered for a second time, MORET issues a warning and fetches the first fit model:

```
iris.rp1<-rpart(Species~.,data=iris)
iris.rp2<-rpart(Species~.,data=iris)
```

```
Note: This model has been created earlier unter the alias "iris.rp1"!
```

6.2 Custom Model Configuration

Model storage and administration tools for the model types *lm*, *glm*, *gam*, *rpart* (see §5.4.1 on page 68) have been supported since the earliest version of MORET. Each type is individually stored in special data base tables. As laid down earlier this kind of design is very hard to adapt to the fast growing number of model packages in R. Special implementation for each package takes too much time. Another solution that is more flexible has been found. §5.4.3.2 on page 81 sketches this flexible solution. Algorithm 3 (p. 83) maps any values from the R object to the generic tree data structure. To remove redundant information a second algorithm must remove unnecessary nodes from this tree. Apart from technical requirements the design of a user interface for the restructuring part proved challenging. The following example illustrates the use of the model configuration wizard.

6.2.1 Example configuration for an rpart tree model

Each model type features common and unique values - and a unique structure. Consequently each type requires a unique mapping. Algorithm 3 maps an arbitrary R object to a generic tree structure. So in the first step the unique model type command must be identified[1]. The R value object to apply algorithm 3 may need extra preparation. Commonly the *summary* report of an R object contains all information required, usually even more than that. The *rpart* type has been chosen because this type does not meet this precondition. No top level statistic for quality and complexity is provided:

```
> names(summary(rpart(Species~.,data=iris)))
...
[1] "frame"    "where"    "call"      "terms"     "cptable"  "splits"
[7] "method"   "parms"    "control"   "functions" "y"        "ordered"
```

Listing 6.1: rpart standard summary

---

[1]See §5.4.3.3, the model must be identified when fitting is done.

The overview table 5.3 (p. 72) relies on these top level statistics. So substitutes must be provided:

```
c(
summary({0}),
splitcount=dim(summary({0})$cptable)[1],
missclassrate=summary({0})$cptable[dim(summary({0})$cptable)[1],
dim(summary({0})$cptable)[2]]
)
```

Listing 6.2: modified rpart summary replacement

splitcount is used as a complexity measure while *missclassrate* - the number of wrongly classified samples is used to measure model quality. The *rpart* model class is available as a hard-coded model type. Still it makes a good example because all configuration features are required. The wizard requires the command for the new type (*rpart*) as well as a valid R-command that produces all relevant data.

Figure 6.2: configuration preparation step: command, name and r-summary command are required for the model tree

An optional description for the model can be added. MORET creates a generic tree object from the R object. The `custom command` above substitutes the standard *summary* for rpart. Algorithm 3 produces a model tree (see figure 6.3). The summary object, created by listing 6.2, contains all information needed. Even more than required. *frame, where, method, params, control, functions* and *ordered* contain information that is not relevant for model selection. Not all of the information contained is irrelevant for model selection. But some is of little value for that purpose. Possibly the *frame* object might replace the *cptable* if more detailed information concerning the splits is desired. For standard comparison the *cptable* is sufficient. In this configuration step the high level statistics for the overview table are selected, one for complexity and one for quality.

Since there were no standard measures the newly introduced *splitcount* and *missclassrate* are used for that purpose. Future model graphs are transformed as configured in this wizard. Any node can be moved, or removed. Whenever the user is satisfied with the resulting graph this configuration is saved. From that time on MORET processes the user-configured command. That means that nodes, identified by name, are removed or moved as defined in configuration. Considering the number of values provided from R the configuration using MORET configuration is very simple and efficient.

① create a model object to extract the values

❶ provide the command

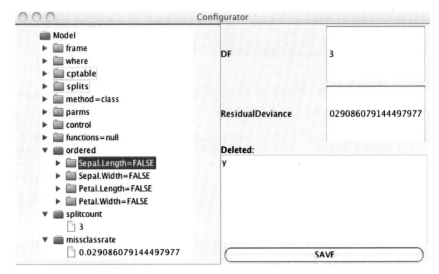

Figure 6.3: default model tree created from wizard in figure 6.2. This default tree contains every value and structure as the replaced summary command. Thus some redundancy needs to be pruned using this wizard.

❷ provide a prototype model command

❸ optionally employ customization for extended/altered summary reports

② drag and drop the global statics for complexity and quality into the named target fields

③ reduce information by filtering information

These steps adapt any new model type to MORET without the need of programming skills.

6.3 Models in MORET

As described in the previous section any available model type can be processed. Hardcoded models may be substituted and customized without great effort. No custom model is required to examine the election data though (see §5.2 on page 52). This example was chosen because a great number of models are available for selection. The original purpose of the data is to find out if the electronic touch screen devices hold an impact on the election result. Evidence is very difficult to come by. No substantial evidence has been found in §5.2 before. The sample distribution does not provide clear support for complex non-linear models. Still, due to the number of variables - increased by potential transformations - a large number of models are found in the model space. Instead of using biased standard forward or backward selection strategies MORET provides a graphical wizard for creating combinations of models for later use as a candidate set.

This section shows an example how to leverage model creation applying MORETs wizards. The used tools - in this example - are explained in detail. But there are too many options to present all of them thoroughly. The R reference manual covers the R-related parameters, the rest can be found in the MORET reference manual (Seger, 2006).

Modeling Approach: First a large number of models is fit using the wizard mentioned above. Additionally a constraining filter is employed. Nonetheless too many models are fit. Thus further interactive search is employed to produce a feasible candidate set.

Model space reduction is performed interactively using the `model explorer` (see figure 6.14 on page 107). The model explorer supports interactive model set selection. From the overview table models are either added or removed. Simultaneously the model explorer extracts model statistics or calculates derived statistic - e.g. Akaike weights (formula 3.3 on page 27). This information is displayed as a value table and graphically. Each derived statistic is computed from a *manually* entered formula. Akaike weights are derived from akaike likelihood which depends on the akaike differences. Thus the easiest way is to calculate the $AIC_{difference}$ first:

☞ Use `add column`, e.g. `custom name` *AICdiff* and `formula ${AIC}-cmin(${AIC})`. `${AIC}` refers to AIC value of each model in the currently selected model set. *cmin* is a specific function that selects the minimum over all candidates.

☞ Since the likelihood depends on this difference the easiest way is to rely on the just intro-duced *AICdiff*. New `custom` *AIClikelihood* is calculated by the `formula exp(-0.5*${@AICdiff})`. The `${@AICdiff}` refers to the new value defined above.

☞ akaike weights are now calculated by the formula
`${@AIClikelihood}/csum(${@AIClikelihood})`

These three steps cover all special cases how to use derived statistics. For convenience *ratio* type statistics can be created in one step, too. More about the *model explorer* is found after figure 6.16 on page 108.

To take advantage of the model explorer a set of models needs created.

6.3.1 How to create a candidate set

Create models employing the model wizard:

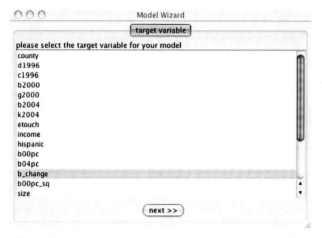

Figure 6.4: model creation wizard step 1: select the target variable

1.

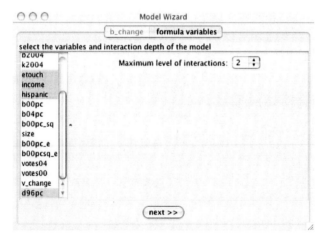

Figure 6.5: model creation wizard step 2: select model variables and interaction depth

2.

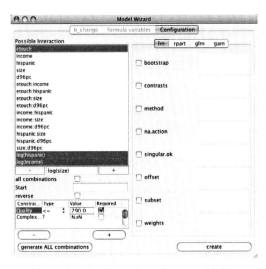

Figure 6.6: model creation wizard step 3: configure model class specific details

3.

Three steps are necessary, to

① chose the target variable

② select the model variables (including interactions)

③ adjust model specific parameters

    ✔ e.g add transformed parameters, splines ...

    ✔ add data specific constraints

    ✔ or *R*-specific parameters

and create one or many model(s) with respect to the manual adjustments.
Figure 6.6 displays all options that MORET supports for specific model creation.

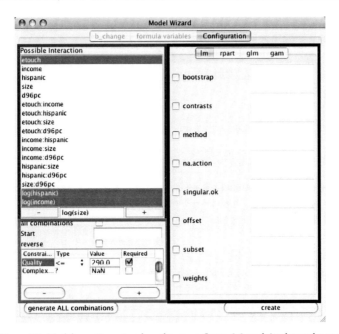

Figure 6.7: Model creation wizard as shown on figure 6.6 explained part by part

blue On the top the desired model type is selected. Below the size for bootstrap samples is determined. Further down all parameters associated with the selected type (R - parameters) are accessible. Each parameter needs to be enabled and assigned a feasible value.

red The left hand side is divided in two parts for variable selection. The upper window contains a list of all variables available for the model. Below additional variables - transformed or derived - may be inserted by hand. Of course variables may be removed. Each selected (highlighted) variable is considered for model generation. These variables are used for combinatoric selection by including (1) or excluding (0) a variable. Since the variables are numbered each combination of ones and zeroes may be represented as an integer.

green This part provides additional control for model creation:

[all combinations] any interaction is allowed for model fitting, no matter if the compound variables are present.

[Start] positive integer representing the start combination. Used to resume work

[reverse] combinatorics set to remove variables, usually the combination starts with $\{0,0,\ldots 0\}$. The number of variables represent the dimension of that vector. Reverse starts with ones, not zeroes.

[constraint table] enforces attribute specific constraints. Models are only stored if they meet all preconditions. E.g. if a certain coefficient must exist enforce this by adding

`Coefficients/{name}/Value`. Also numeric constraints are supported, like $AIC \leq$ *threshold* .

Instead of using classical forward or backward model selection with an arbitrary step and stopping rule the whole model space is scanned for valid models.

6.3.2 Constrained model Creation

Why are constrains a good idea for model creation ? Remember the problems encountered when the election data set was scanned for models. Five base variables and ten two-way-interactions produce 32767 models. The number of good models is much smaller. Quality based model inference removes globally bad models from a candidate set, but why should *bad* models be regarded at all? Using an approximate model as a reference goodness of fit and a threshold for eligible models reduces the exhaustive set to a manageable one.

As a rule of thumb 5 percent difference between the reference model yields a manageable number of models.

```
> AIC(lm.fl04.best)
[1] -321.7194
> AIC(lm.fl04.full)
[1] -295.0725
```

As first reference threshold the full model $AIC \cdot 1.05$ is used. $AIC_{full} = -295.0725$. $AIC_{ref} \leq -309.8261$ (see figure 6.6 on page 97 for the creation screenshot).

Using this constraint 8467 models still remain, that feature a better AIC than $AIC_{reference}$. That is 25.8% of the exhaustive model set and given a tool like MORET such a number is *manageable*.

A narrower constraint like $AIC \leq 1.01 \cdot AIC_{Best}$ still leaves 2362 or 7.2% of the exhaustive set. If the set of models is still too large, additional constraints need to be applied. Such constraints must be chosen according to the problem at hand.

One standard constraint is generally recommended: The compound variables (*base vars*) must be present when interactions are used (by default *all combinations* are disabled). Activating this constraint drops the maximum number of available models to 570 models at the 5% constraint.

Constraint	AIC_{ref}	Threshold	AIC_{best}	# $\leq$ Threshold
-	-295.0725	-309.8261	-321.7194	8467
base vars	-295.0725	-309.8261	-318.3426	570
no *etouch*	-295.0725	-309.8261	-319.8794	256
base vars AND ~~etouch~~	-295.0725	-309.8261	-318.3426	40

Table 6.1: model space reduction by constraint(s)

Table 6.1 shows the reduction induced by constraints. Figure 6.8 displays the best models that remain when *base vars* are mandatory: No model in this set could best lm.fl04.best, based on AIC quality. *Backward variable selection* chose the same best model with respect to the AIC. The candidate set (figure 6.8) was created using a threshold value of $AIC \leq 1.05 \cdot AIC_{Best}$. In

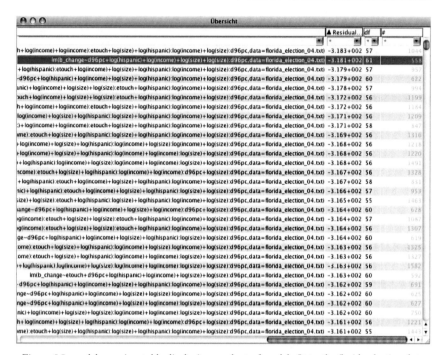

Figure 6.8: model overview table displaying a subset of models fit to the florida election data set, ordered by best quality

this set of 8467 models 8211 models contain some *etouch* interaction. Conversely 256 models do not support *etouch*. The number of models that support a variable is no reasonable selection criterion. When competitive parsimonious models exist this variable might be dropped, too.
As a contrast to the very complex lm.fl04.best the model with the lowest complexity in this set should be examined:

```
Call:
lm(formula = b_change ~ d96pc + log(size):d96pc + log(size),
    data = florida_election_04.txt)

Residuals:
      Min         1Q      Median        3Q        Max
-0.0526211  -0.0104128  0.0006586  0.0138769  0.0572057

Coefficients:
             Estimate Std. Error t value Pr(>|t|)
(Intercept)  -0.45511    0.14397  -3.161 0.002417 **
d96pc         1.20941    0.29289   4.129 0.000109 ***
```

```
log(size)           0.04156      0.01280    3.246 0.001876 **
d96pc:log(size)    -0.10456      0.02615   -3.998 0.000170 ***
---
Signif. codes:  0 *** 0.001 ** 0.01 * 0.05 . 0.1   1

Residual standard error: 0.02287 on 63 degrees of freedom
Multiple R-squared: 0.4036,     Adjusted R-squared: 0.3752
F-statistic: 14.21 on 3 and 63 DF,  p-value: 3.541e-07

> AIC(fl.df63)
[1] -310.2039
```

Since lots of models have been found that satisfy the chosen constraint, the best model that does not include *etouch* is compared against the model *fl.df63*:

```
Call:
lm(formula = b_change ~ d96pc + log(hispanic) +
   log(income):log(hispanic)  + log(size):d96pc + log(size))

Residuals:
      Min         1Q      Median        3Q         Max
-0.0514884 -0.0133498  0.0002598  0.0115638  0.0405599

Coefficients:
                           Estimate Std. Error t value Pr(>|t|)
(Intercept)               -0.588340   0.136937  -4.296 6.34e-05 ***
d96pc                      1.317070   0.270338   4.872 8.21e-06 ***
log(hispanic)             -0.227754   0.074646  -3.051  0.00337 **
log(size)                  0.050211   0.011984   4.190 9.14e-05 ***
log(hispanic):log(income)  0.021163   0.007161   2.955  0.00444 **
d96pc:log(size)           -0.110570   0.024120  -4.584 2.31e-05 ***
---
Signif. codes:  0 *** 0.001 ** 0.01 * 0.05 . 0.1   1

Residual standard error: 0.02099 on 61 degrees of freedom
Multiple R-squared: 0.5137,     Adjusted R-squared: 0.4738
F-statistic: 12.89 on 5 and 61 DF,  p-value: 1.463e-08

> AIC(lm.noeto.best)
[1] -319.8794
```

The AIC_{weight} is related to the likelihood to select a model in a candidate set (see formula 3.3). In this competition the likelihood of this *lm.noeto.best* of representing the best model is 16.5%.

The gain in quality is small, still the R^2_{adj} is a bit better. To decide if the *etouch* term should be included for structural purposes some of the default analytical model plots from *R* should be considered:

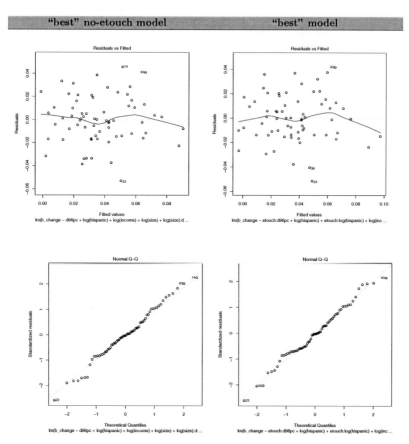

Figure 6.9: Selection of R's default plots for linear models. The best model with and without *etouch* are compared

6.3.3 Summary

Since the quality difference between those two models is small - and even the analytical plots look fine - the no- *etouch* model is a good choice to use for inference. Second order interactions were excluded from model selection because these are more difficult to interpret than first-order. No *etouch* coefficient remains in the selected model - still analysis at this point is not satisfactory and is continued in the next section (from page 103 on).

More sophisticated methods than OLS exist to fit models. The selected models were displayed to demonstrate the amount of data encountered in a straightforward example. No variable transformations were necessary or complex model types. Nonetheless no clear result was identified so far. Rather a set of models is found. Sophisticated variable selection methods like Bayesian methods (§3.3.6 on page 36) are helpful for finer model space selection. But this is a different approach.

The main focus was to show how to create sets of models and rank these on a practical example. These models may be ranked on a global level, like *df*, *RSS*, *AIC*, *deviance*, ... Low level statistics cover model quality, complexity or a combination of both. For the linear models above the best known global statistics are *df* the degrees of freedom for model complexity and R^2 as a global measure of fitness. R^2_{adj} is a penalized global statistic. Alternatively Mallows C_p (R package mle), *AIC* (stats) or *deviance* may be used. Globally bad models are usually not be considered for candidate sets - that's why the threshold value was chosen to reduce the set of 32767 models to a manageable number.

On a more detailed level the model coefficients come into account. Preferably only explainable coefficients and interactions are selected. If a model or set of models is used for inference n-way (n > 2) interactions should be avoided. Ignoring any of these recommendations an even larger set of models has to be considered and evaluated.

At the most detailed level the model can be assessed on the residuals (or fit versus observed values). Here the model shows how well it fits the particular data set. Residual patterns differ from model to model and anomalies are hard to identify. That holds especially true for subset comparison. Usually the term outlier is used if single or few observations play a dominating role in the model. Outliers need to be treated carefully and often the removal of an outlier is preferable. In particular if no sound model has been found offending outliers should be removed. If any observation is classified an error this observation must be removed, too.

6.3.4 Typical steps

Regardless whether models are generated using the wizard or manually, MORET stores all of these models in its database for further processing. The next step is to inspect the generated models. Assuming that there have been a number of data sets analyzed the first step is always to select a subset of all stored models (figure 6.10): Figure 6.10 lists all models from the database where the target variable b_{change} has been used from the florida election data set and the type is either a linear or generalized linear model.

The resulting table contains all created models - subject to the selected constraints. This table contains too many models to produce a candidate set directly: To reduce the number of available models in this table column-sensitive filters can be applied: in this example only general linear models are considered. The R model class GLM feature AIC. The overview table displays only *residual deviance* at this detail level without a hint what statistic is used (in this table R's `deviance` is used). AIC is useful for later steps. The table entries are sorted ascending with respect to the global quality statistic (figure 6.12):

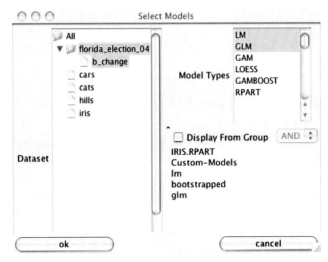

Figure 6.10: standard selection dialog used to pick a valid subset of models for the overview table

File	Target Va...	Type	User Name	Residual Err...	df	R Command
la_election_04.txt	*	*	*	*	*	*
florida_election_04.txt	b_change	LM	lm.1	+3.737E-002	+6.200E+001	lmlb_change~d96pc+hispanic+income+e
florida_election_04.txt	b_change	LM	lm.2	+2.907E-002	+5.100E+001	lmlb_change~d96pc*hispanic*income*etc
florida_election_04.txt	b_change	LM	florida_election_04.txt.lm.2	+5.211E-002	+6.500E+001	lmlb_change~etouch,data=florida_electio
florida_election_04.txt	b_change	LM	florida_election_04.txt.lm.3	+4.542E-002	+6.500E+001	lmlb_change~income,data=florida_electio
florida_election_04.txt	b_change	LM	florida_election_04.txt.lm.4	+5.145E-002	+6.500E+001	lmlb_change~hispanic,data=florida_electi
florida_election_04.txt	b_change	LM	florida_election_04.txt.lm.5	+5.381E-002	+6.500E+001	lmlb_change~d96pc,data=florida_election
florida_election_04.txt	b_change	LM	florida_election_04.txt.lm.6	+4.481E-002	+6.400E+001	lmlb_change~etouch+income,data=florid.
florida_election_04.txt	b_change	LM	florida_election_04.txt.lm.7	+4.944E-002	+6.400E+001	lmlb_change~etouch+hispanic,data=flori
florida_election_04.txt	b_change	LM	florida_election_04.txt.lm.8	+5.105E-002	+6.400E+001	lmlb_change~etouch+d96pc,data=florida
florida_election_04.txt	b_change	LM	florida_election_04.txt.lm.9	+4.207E-002	+6.400E+001	lmlb_change~income+hispanic,data=flori
florida_election_04.txt	b_change	LM	florida_election_04.txt.lm.10	+3.895E-002	+6.400E+001	lmlb_change~income+d96pc,data=florid.
florida_election_04.txt	b_change	LM	florida_election_04.txt.lm.11	+5.078E-002	+6.400E+001	lmlb_change~hispanic+d96pc,data=floric
florida_election_04.txt	b_change	LM	florida_election_04.txt.lm.12	+4.189E-002	+6.400E+001	lmlb_change~etouch+income+hispanic,d.
florida_election_04.txt	b_change	LM	florida_election_04.txt.lm.13	+3.894E-002	+6.300E+001	lmlb_change~etouch+income+d96pc,dat
florida_election_04.txt	b_change	LM	florida_election_04.txt.lm.14	+4.891E-002	+6.300E+001	lmlb_change~etouch+hispanic+d96pc,da
florida_election_04.txt	b_change	LM	florida_election_04.txt.lm.15	+3.737E-002	+6.300E+001	lmlb_change~income+hispanic+d96pc,d.
florida_election_04.txt	b_change	LM	florida_election_04.txt.lm.16	+3.737E-002	+6.200E+001	lmlb_change~etouch+income+hispanic+c
florida_election_04.txt	b_change	GLM	florida_election_04.txt.glm.2	+5.211E-002	+6.500E+001	glmlb_change~etouch,data=florida_electi
florida_election_04.txt	b_change	GLM	florida_election_04.txt.glm.3	+4.542E-002	+6.500E+001	glmlb_change~income,data=florida_electi
florida_election_04.txt	b_change	GLM	florida_election_04.txt.glm.4	+5.145E-002	+6.500E+001	glmlb_change~hispanic,data=florida_elec
florida_election_04.txt	b_change	GLM	florida_election_04.txt.glm.5	+5.381E-002	+6.500E+001	glmlb_change~d96pc,data=florida_electio
florida_election_04.txt	b_change	GLM	florida_election_04.txt.glm.6	+4.481E-002	+6.400E+001	glmlb_change~etouch+income,data=flori
florida_election_04.txt	b_change	GLM	florida_election_04.txt.glm.7	+4.944E-002	+6.400E+001	glmlb_change~etouch+hispanic,data=flor
florida_election_04.txt	b_change	GLM	florida_election_04.txt.glm.8	+5.105E-002	+6.400E+001	glmlb_change~etouch+d96pc,data=florida
florida_election_04.txt	b_change	GLM	florida_election_04.txt.glm.9	+4.207E-002	+6.400E+001	glmlb_change~income+hispanic,data=flo
florida_election_04.txt	b_change	GLM	florida_election_04.txt.glm.10	+3.895E-002	+6.400E+001	glmlb_change~income+d96pc,data=flori
florida_election_04.txt	b_change	GLM	florida_election_04.txt.glm.11	+5.078E-002	+6.400E+001	glmlb_change~hispanic+d96pc,data=flor
florida_election_04.txt	b_change	GLM	florida_election_04.txt.glm.12	+4.189E-002	+6.300E+001	glmlb_change~etouch+income+hispanic,d
florida_election_04.txt	b_change	GLM	florida_election_04.txt.glm.13	+3.894E-002	+6.300E+001	glmlb_change~etouch+income+d96pc,da
florida_election_04.txt	b_change	GLM	florida_election_04.txt.glm.14	+4.891E-002	+6.300E+001	glmlb_change~etouch+hispanic+d96pc.d
florida_election_04.txt	b_change	GLM	florida_election_04.txt.glm.15	+3.737E-002	+6.300E+001	glmlb_change~income+hispanic+d96pc.

Figure 6.11: the model overview table selected using the dialog (figure 6.10)

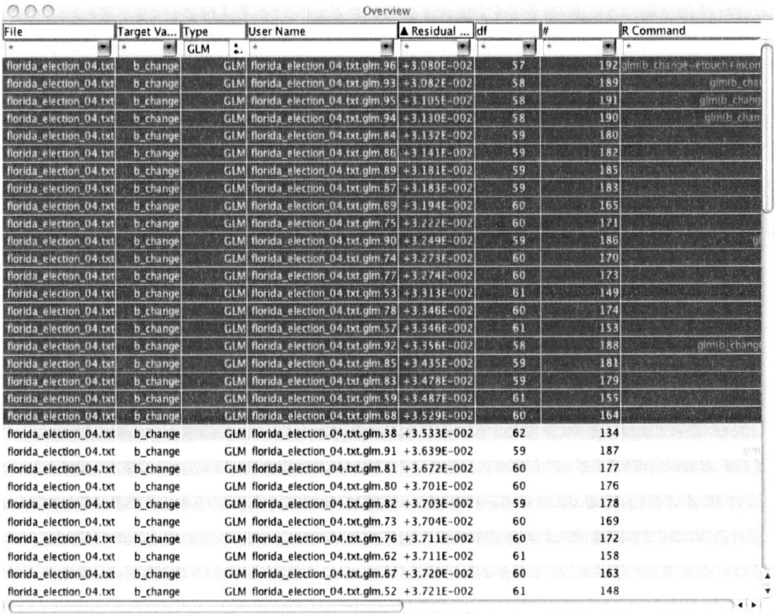

Figure 6.12: initially the set of 21 models is selected in the model overview table for detailed investigation

More sophisticated subselection is achieved using regular expressions, e.g. to filter out variables, in combination with sort order. The selection from this table provides the models for detailed evaluation. Usually candidate sets are picked by the best quality measure (figure 6.12). The resulting set of models is inspected at intermediate level of detail. Any structural attribute can be

✔ selected,

✔ sorted or

✔ exported

for further processing with external advanced tools (figure 6.13). This example selection is sorted on the p-value of *etouch* and global quality statistics are selected for each model. The number of selected intermediate statistics creates a wide table, thus only a part can be displayed on one image. Any selected values can easily be transferred to a spreadsheet program using copy and paste or drag and drop. Advanced analysis is easily accessible either this way or by data export.

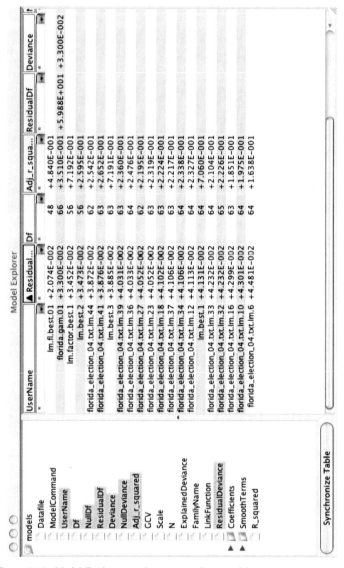

Figure 6.13: Model Explorer: analyze intermediate model statistics interactively

Different types of models may be evaluated within the same explorer table. This example has been created with a specially modified version of the linear model[2]. For better visual distinction deviance has been spelled in lowercase letters on figure 6.14. *AIC* is featured in both model types so this statistic can easily be compared along different model types.

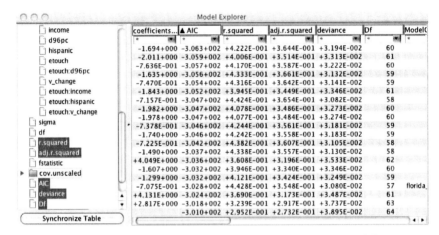

Figure 6.14: Model Explorer: only linear model attributes are selected for detailed comparison

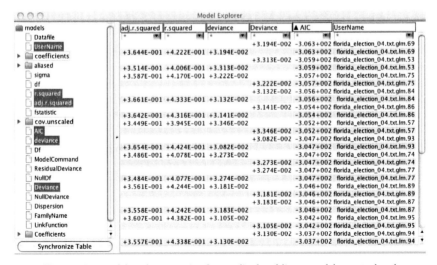

Figure 6.15: model explorer: a mix of generalized and linear models are analyzed

---

[2]How to substitute a model is described in §5.4.3.3 on page 82.

The linear models feature the same AIC values as the generalized type. In this example no *glm* is subject to a categorical link function. That means the link function is "identity" and family is "gaussian". So *lm* and *glm* cannot be distinguished in terms of prediction. As a result only the *glm* subset is considered for candidate set selection. All of these values have been transferred to the spreadsheet program. Multi model statistics are computed in this spreadsheet program semi-manually[3], in this example (cf. figure 6.16). MORET offers possibilities in the *model explorer* to compute these statistics automatically and subject to the present model subset. (see section 6.5 on page 118)

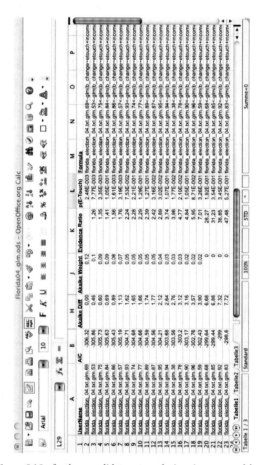

Figure 6.16: further candidate set analysis using a spreadsheet

---

[3]of course the formulae are copied

The single best selected model is not convincingly superior to the other good models. Using the AIC interpretation for the current set, it takes 6 models to exceed 50 percent support, 11 to best 75 percent and 18 models for 99 percent. No satisfying conclusion can be drawn by multi model inference, yet. The original question if electronic touch screen devices have an impact on the election result cannot be answered. The data set features a lot of variation and instability - combining evidence weight and *etouch* factor probabilities yields mediocre support for an *etouch* effect. Another interactive approach will be given later on (see section 6.5 on page 118).

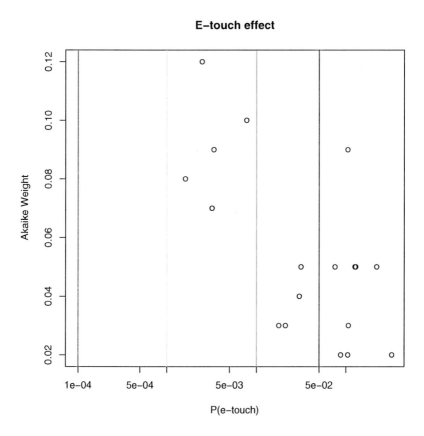

Figure 6.17: *etouch* p-values vs. AIC for the candidate model set

6.3.5 Feature Queries - An alternative model table selection approach

Managing model statistics is one task MORET has been created for. From the model table (figure 6.11 on page 104) only few attributes of models are visible for pre-selection. The models are selected from this table (figure 6.11) which provides manual filter tools. Any table content was pre-filtered by a selection dialog 6.10 (page 104). Though this kind of selection is appropriate and later filtering useful there are some drawbacks for this kind of approach. The model overview table is chosen exclusively by the dependency to data, variable selection and group assignment. Though group assignment can be the most useful selection method, a group must be assembled first. Doing so manually or using the model explorer (figure 6.13, page 106) is a laborious task. The model overview table as described above, does not support selection on model attributes or values. This overview table is the source for many other steps like group assignment. Furthermore model set selection, as displayed in figure 6.16, is a typical task where a software tool requires manual supervision. Directly selecting models by special model statistics is another tool, since this introduces new means to select candidate sets. Instead of using evidence ratio for model selection, models can be selected by an arbitrary combination of features - like "meaningful coefficient". "Meaningful coefficient" is common speech that needs to be translated into a language, a computer can process, to provide the desired data. The theoretical background has been introduced in §5.4.4.1 on page 84.

To fulfill that request MORET offers either direct SQL input for an expert user or a graphical tool. This graphical tool does not require the user to know about SQL syntax and thus frees the user from the burden of learning the data schema. Instead each model statistic, or *"feature"*, can be used in combination with other features.

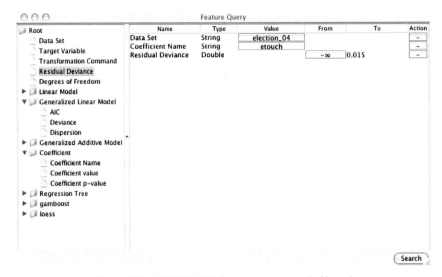

Figure 6.18: MORET GUI: feature query example (Query)

```
Keep in mind that the resulting table displays only those models that
contain every feature.
```
The aim is to find out if there are models in the *florida elections* data set that feature a coefficient containing *etouch* and the global quality statistic is below 0.015 (figure 6.18).

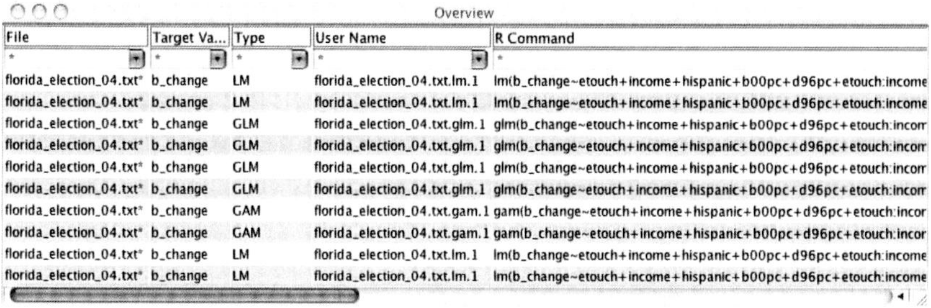

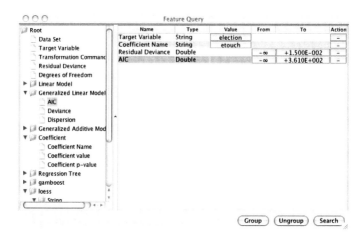

Figure 6.19: MORET GUI: feature query example result: an overview table

This provides a very versatile way of selecting model sets - one thing should be kept in mind: Some features are specific to one model type and selecting these features implicitly restricts the results to that particular type.

Figure 6.20: MORET GUI: feature query example 2 (Query)

Figure 6.21: MORET GUI: result of feature query example 2 (Figure 6.20)

Even more sophisticated (combined) queries are possible. An empty result may unintentionally come up due to the selection of opposed features. If the models set to be found should contain one coefficient named *etouch* with a maximum $p - value < 0.01$ and another coefficient named *income* with the maximum $p - value < 0.01$ these features need to be "grouped". That means the feature is not only in the same model but the same attribute (i.e. the same parent node).

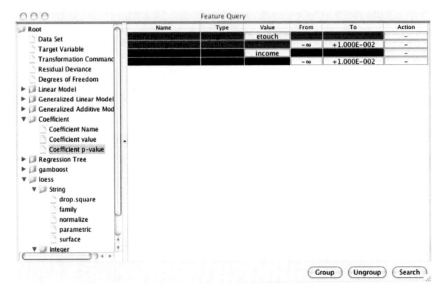

Figure 6.22: MORET GUI: combined feature query example (Query). Two named coefficients that feature a p-value lower than 0.01 are searched for.

Figure 6.23: combined feature query example (Result Details)

Without the combination the resulting query has the following form :

```
SELECT DISTINCT arg_0.id
FROM coefficients arg_0
JOIN coefficients arg_1 ON arg_0.id = arg_1.id
JOIN coefficients arg_2 ON arg_1.id = arg_2.id
JOIN coefficients arg_3 ON arg_2.id = arg_3.id
WHERE lower(arg_0.name) like '%etouch%'
AND arg_1.p>'-Infinity' AND arg_1.p<0.01
AND lower(arg_2.name) like '%income%'
AND arg_3.p>'-Infinity' AND arg_3.p<0.01
```

This query produces obviously more hits than the AND combination:

```
SELECT DISTINCT grp_000.id
FROM coefficients grp_000
JOIN coefficients grp_001 ON grp_000.id = grp_001.id
WHERE lower(grp_000.name) like '%etouch%'
AND grp_000.p>'-Infinity'  AND grp_000.p<0.01
AND lower(grp_001.name) like '%income%'
AND grp_001.p>'-Infinity'  AND grp_001.p<0.01
```

The problem is not writing this kind of query but to provide a user-friendly tool for just anyone who is not familiar querying a database using SQL. Selecting features is a convenient way to chose model sets. Since the features data-type is either numeric or a string the input is supervised from the tool. That prevents many empty results. Additionally combining (grouping) features is another useful way to control model set selection.

6.4 Organizing Data

So far the process how data from R is stored into a relational database has been explained thoroughly. The models are stored in a database and can be presented neatly as a table containing all useful information. See table 5.3 on page 72 for the concept or figure 6.8 on page 100 for a practical example.

The set of models can be explored from this table using filters and sorting. The gain in model set selection is remarkable, compared to manual techniques or simpler computer aided specialized tools. *Specialized* means that this tool only solves one, or few of the many tasks related to multi-model management. Unfortunately the previously discussed techniques do not suffice for really large model sets. Especially bootstrapping inflates the model space and as a result the model table becomes confusing.

Hand-picking model sets from a long list is tedious, considering MORET uses a database internally. This database provides SQL for exact model finding. When the analyst is familiar with SQL model sets can be picked effectively using model values. Otherwise alternative means must be found, to reduce the enormous amount of data. Table sorting on attributes helps as well as fine-grained table preparation and management.

Nonetheless this is still laborious. So MORET offers additional tools to facilitate this task for the user.

6.4.1 Filters

The first way to cope with long lists of models is to use the filters from MORET. In fact filters have been used earlier (Figure 6.12, page 105) but not formally introduced. Filtering is a straightforward approach when the amount of data becomes too large, restrict the data with respect to a certain constraint - a *filter*.

Whenever a user looks for models that have been stored inside the database a preselection can reduce the enormous amount of information to a manageable set. Preselection can be performed as displayed on (Figure 6.10, page 104) or using the feature query facility (§6.3.5, p. 110). Preselection is the first step to select the model overview table. After that the table content is reduced by filters and sorting to provide an effective means for model selection.

Preselection Filter The first preselection filter criterion is the data set and the selected target variable. That usually picks more than a candidate set but fortunately not all other data sets or undesired target variables interfere with the search. The next preselection filter is the model class. Possibly many model classes are tried out while picking the candidate set. Not all if these classes are found suitable later. Getting rid of unsuitable models is important to form candidate sets. In the decision phase - until the knowledge has been gained what models are suitable - such models are stored in the database. Until these models are deleted a the preselection result is unclear.

To allow finer model preselection *groups*, that is a logical unit models can belong to, are introduced in §6.4.2 (p. 116). Assigning models to groups and later using exclusive or combined group assignments in the preselection filter imposes an effective restriction for better candidate set model selection.

This preselection results in a list of every model created that matches the above filter criteria.

Model Table Filters The combo-boxes provide an easy way to filter actively for a manageable set of models. When the preselection result contains too many hits the model table offers filter criteria on the top row. An expert user can even use regular expressions (see App. §B.3 on 157) as a filter criterion.

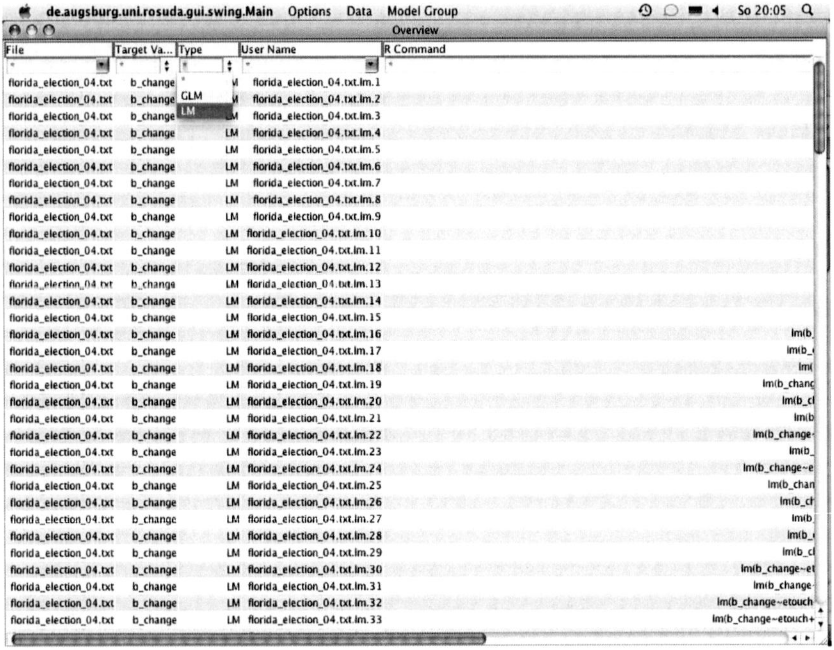

Figure 6.24: model table filters reduce the number of displayed models in the overview table
interactively

6.4.2 Groups

Models belong to the data set they were created from, this includes sub sampling like bootstrapping techniques. Additionally models might even share other types of interrelationship that are not straightforward. To specify such a model interrelationship these models can be assigned to an arbitrary *group*. This *group* is a logical unit, described a name and an optional description (see 6.4.2 on p. 116).

Creating groups and assigning or removing models to these groups is a manual effort. As mentioned above there are interrelationships that can not be logically derived. This administrative effort is rewarded by a finer grained model preselection (6.4.1, page 115). Figure 6.10 on page 104 shows the preselection dialog - with a number of groups in the right bottom compartment. Group names like `lm` or `glm` are not very helpful. A a better name is `aic-best 20 linear models` or another specific name.

The concept of groups provides effective combinations with the earlier mentioned preselection features (see right lower part of figure 6.10, p. 104). Using `AND` the resulting models are contained in all of the selected groups. `OR` allows any model that belongs to one of the selected groups. Thus groups provide better control over the resulting model table. From this table other administrative tasks are performed. Any selected set of models can be added to or removed from a group. Current group members contained in this table can be selected or groups (not the assigned models) can be deleted or created:

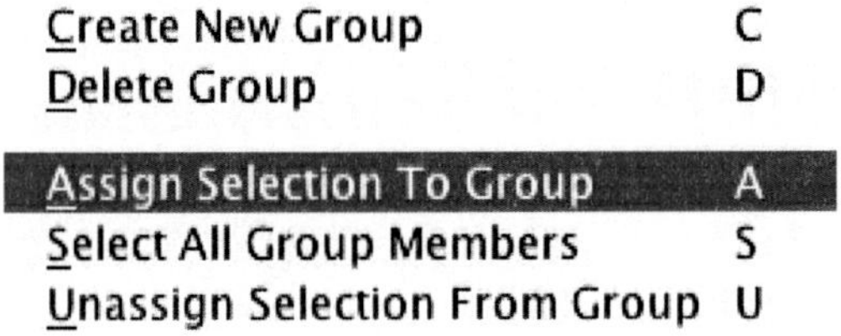

Create New Group	**C**
Delete Group	**D**
Assign Selection To Group	**A**
Select All Group Members	**S**
Unassign Selection From Group	**U**

Figure 6.25: Model Overview Table Menu: Group Management

6.5 Example: Interactive Search

6.5.1 Florida Election Data

This data set was discussed in detail in §5.2. This time a large number of models will be created using the internal model generator. Thereafter interactive tools are used to check if there is support for models containing *etouch*. The models are generated from *income,hispanic*, v_{change}, *d96pc, etouch* and all two-way interactions. Five variables allow ten two-way interactions. When interactions are restricted by the availability of compound variables the model space is 1449 out of 32768. In this set of models all coefficient probabilities are highlighted successively from the better models first to the not-so-good models.

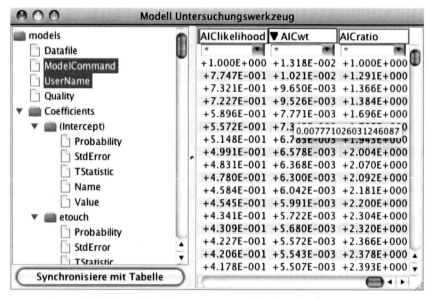

Figure 6.26: The model explorer supports easy Akaike weights[4] computation

Figure 6.26 shows that, according to the Akaike weights and likelihood ratios, there is no single dominant model. Using Akaike for selection (cf. figure 6.26)

---

[4]Whenever the mouse pointer hovers over a table cell the value is shown in a yellow overlay in maximum detail level

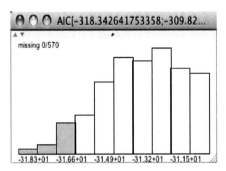

Figure 6.27: MORET GUI: select Akaike weights according to the selected models models in figure (6.26) as histogram

the coefficient probabilities are scanned for instances of models that feature "significant" *etouch* probabilities.

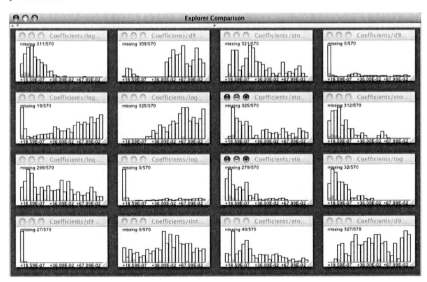

Figure 6.28: MORET GUI: interactive histograms of coefficient/p-value

Examining these histograms (Figure 6.28) "significant" *etouch* coefficients are encountered. The presence of *etouch* in this set is not surprising. The other way round is more interesting: Are there good models without *etouch* as well ? A simple way to select no-*etouch* models is to use the

model explorer as shown in Figure 6.26. Select the *etouch*-coefficient and sort on that column.
Any cell that holds no value does not contain *etouch*.

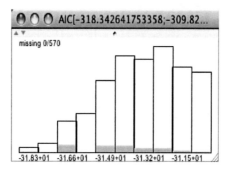

Figure 6.29: Distribution of Akaike weights as histogram, no-*etouch* models are highlighted

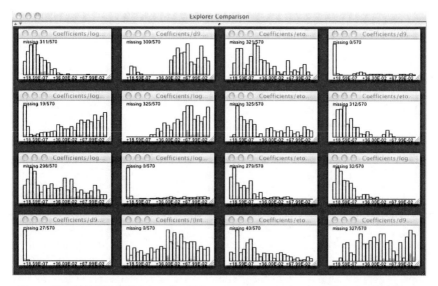

Figure 6.30: MORET GUI: interactive histograms where no e-touch coefficient probability is
highlighted

Selecting no-*etouch* coefficients (cf. figure 6.29, p. 120) shows that few of the best models contain
an *etouch* coefficient. So the idea that was already mentioned in §6.3.3 on page 102 reemerges:
the *etouch* effect is negligible.

An alternate interactive tool

Figure 6.29 on page 120 showed that no-*etouch* models are competitive with the models involving this additional variable. The facilities of MORET regarding interactive data analysis depends mostly on third party software. Mondrian (Theus, 2002b) is a tool that provides many ways of interactive data selection. From MORET's overview table, as displayed on figure 6.8 on page 100, the data may be exported in a format readable by Mondrian. There a "missing value plot" is created and all no-etouch models are selected.

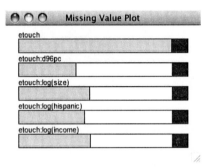

Figure 6.31: Mondrian: missing value plot - no *etouch*

Now the model quality is tested against these remaining models:

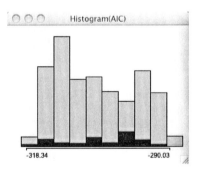

Figure 6.32: Mondrian: AIC values, where no e-touch is selected (Figure 6.31), same as 6.29

Figure 6.32 shows that omitting *etouch* leaves a subset of good models. Considering model complexity and sparse model design *etouch* can be omitted from modeling as pointed out before. Using the abilities of model management from MORET, generation, storage and interface output, combined with interactive model selection like Mondrian new capabilities arise.

6.5.2 Bodyfat Data

The Florida elections data set featured a lot of variables. Many variables were directly derived though. Another data set that features a lot of highly correlated variables, has been collected to predict human body fat from a number of different measures.

Figure 6.33: Measuring Body Fat [5]

These   measures   do   not   only   cover   weight   but   feature   the   following   variables

- Age (years)
- Weight (lbs)
- Height (inches)
- Neck circumference (cm)
- Chest circumference (cm)
- Abdomen circumference (cm)
- Hip circumference (cm)
- Thigh circumference (cm)
- Knee circumference (cm)
- Ankle circumference (cm)
- Biceps (extended) circumference (cm)
- Forearm circumference (cm)
- Wrist circumference

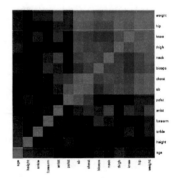

Figure 6.34: bodyfat correlation heatmap

This data set can be found in Behnke and Wilmore (1974), pp. 45-48 which covers additional details on how measures were taken. The heat map (cf. figure 6.5.2) shows that most variables are highly correlated - full green color stands for positive, red stands for negative correlation.

13 different variables combine to 8192 different models - without interaction or transformation of variables. Combining these with two way interactions provides 2^{91} different possible models (when no compound variable constraint is imposed).

Different approaches have been tried to find good explanatory models.

Burnham and Anderson (2010) created derived variables (see page 268); other modeling approaches are also mentioned. In the 252 samples outliers can be identified (see figure 6.35). Two data sets of models are analyzed with MORET, one containing all observations and one without the outliers. Any observation where weight is located beyond the whiskers[6] is considered an outlier.

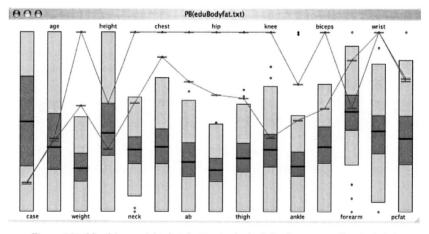

Figure 6.35: Mondrian: weight distribution in the bodyfat data set, outliers included

6.5.2.1 Using two concurrent data sets

The variables are highly correlated. Thus lots of models with little quality difference are to be expected. The first step is to calculate the best linear model to produce a threshold quality. To check how much weight each variable contributes in the models all variables are scaled to a range from zero to one. Employing R's *stepAIC* method we end up with a best model (using the "cleaned up"[7] data set)

---

[6]MONDRIAN plots a light gray colored solid box as whisker
[7]The two observations with the largest weight have been removed. See figure 6.35

```
Call:
lm(formula = pcfat ~ age + height + neck + chest + ab + forearm +
    wrist, data = bfatNormalizedNoOutliers)


Residuals:
      Min        1Q    Median        3Q       Max
-0.220231 -0.066253 -0.005605  0.063951  0.198691


Coefficients:
             Estimate Std. Error t value Pr(>|t|)
(Intercept)   0.11408    0.17195   0.663  0.50768
age           0.09624    0.04263   2.258  0.02485 *
height       -0.56765    0.19989  -2.840  0.00490 **
neck         -0.32981    0.20284  -1.626  0.10526
chest        -0.32896    0.22421  -1.467  0.14362
ab            2.25797    0.16666  13.548  < 2e-16 ***
forearm       0.25924    0.14090   1.840  0.06701 .
wrist        -0.70541    0.21738  -3.245  0.00134 **
---
Signif. codes:  0 '***' 0.001 '**' 0.01 '*' 0.05 '.' 0.1 ' ' 1

Residual standard error: 0.08906 on 242 degrees of freedom
Multiple R-squared: 0.7465,     Adjusted R-squared: 0.7392
F-statistic: 101.8 on 7 and 242 DF,  p-value: < 2.2e-16
```

The AIC for this model, fit to the data set without outliers, is -489.91. A threshold level of one percent will filter out a lot of bad models. This threshold[8] is used for both data sets. For better visibility the plots are presented side by side: No dominant best model can be identified which is no surprise, considering the high correlation. The histograms can be interactively modified, e.g. the number of bins and the current selection. Mondrian supports the evaluation of the selection frequency of any variable interactively, e.g. by selecting the best of the best from the AIC histogram. The coefficient value and p-value can be visualized concurrently. Figure 6.36 shows the selection of the very best models concurrently with parallel boxplots of the coefficient values and p-values. Some test based selection algorithms use p-values for model selection. High p-values (i.e. near one) will usually[9] be dropped in the next iteration step. The model coefficient values are used for forecast purposes. So near null values will not have much effect on any forecast result. The variables order is based on coefficient values for the complete data set. The numerical values are given in table 6.2.

---

[8] The threshold level for the data set, including outliers, is -486.81
[9] p-values higher than 0.1 are considered high, but this example only relies on pure AIC bases selection.

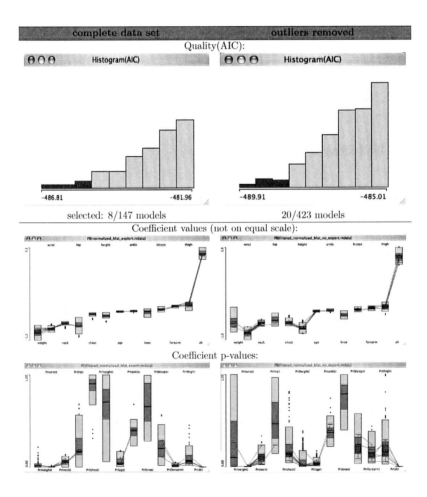

Figure 6.36: Mondrian: model comparison fit to the full data set and the cleaned data set

NAME	all data		outliers removed	
	min	**max**	**min**	**max**
ab	2.669	3.260	1.877	2.565
age	0.066	0.148	0.066	0.143
ankle	0.063	0.186	0.048	0.189
biceps	0.091	0.356	0.072	0.278
chest	-0.262	0.062	-0.455	-0.058
forearm	0.277	0.425	0.112	0.301
height	-0.394	0.072	-0.751	-0.239
hip	-1.122	-0.159	-0.467	0.046
knee	-0.166	0.245	-0.204	0.113
neck	-0.709	-0.362	-0.466	-0.195
thigh	-0.154	0.570	-0.033	0.451
weight	-1.280	-0.552	-0.696	0.165
wrist	-0.898	-0.489	-0.983	-0.485

Table 6.2: coefficient ranges in the normalized bodyfat linear models

From this table variables are chosen by their contributed weight in any model. These variables provide a sub-sample of all possible two-way interactions, Thus a feasible number of models is scanned for additional better models. *ab*, *weight*, *wrist*, *height*, *neck* and *thigh* are chosen. For 6 variables 15 interactions are supported with a maximum model space of 2^{21}.
The best model provides a new threshold quality value similar to before:

☞ create the most complete model command using the MORET model wizard:

```
bfatNormalized.lm.148<-(weight+height+neck+ab+thigh+wrist)^2,
   data=bfatNormalized)
```

☞ use R to find a threshold value

```
> stepAIC(bfatNormalized.lm.148)
...

Call:
lm(formula = pcfat ~ weight + height + neck + ab + thigh + wrist +
    weight:ab + weight:wrist + height:neck + height:ab + neck:ab +
    thigh:wrist, data = bfatNormalized)

Residuals:
      Min        1Q    Median        3Q       Max
-0.232565 -0.055079 -0.001209  0.058815  0.215299

Coefficients:
            Estimate Std. Error t value Pr(>|t|)
(Intercept)   -7.900      3.481  -2.270  0.02412 *
weight        -3.299      3.200  -1.031  0.30371
height         5.649      3.375   1.674  0.09545 .
```

```
neck           7.114    5.847    1.217   0.22491
ab            -5.938    4.119   -1.441   0.15076
thigh         11.344    3.791    2.992   0.00306 **
wrist          4.275    2.013    2.123   0.03477 *
weight:ab     -6.070    1.977   -3.071   0.00238 **
weight:wrist   8.229    3.583    2.297   0.02248 *
height:neck  -13.797    6.342   -2.175   0.03058 *
height:ab      6.244    3.660    1.706   0.08928 .
neck:ab        7.981    3.624    2.202   0.02861 *
thigh:wrist  -13.202    4.422   -2.985   0.00313 **
---
Signif. codes:  0 '***' 0.001 '**' 0.01 '*' 0.05 '.' 0.1 ' ' 1

Residual standard error: 0.08708 on 239 degrees of freedom
Multiple R-squared: 0.7674,    Adjusted R-squared: 0.7557
F-statistic: 65.71 on 12 and 239 DF,  p-value: < 2.2e-16

> AIC(lm(formula = pcfat ~ weight + height + neck + ab + thigh +
+ wrist + weight:ab + weight:wrist + height:neck + height:ab +
+ neck:ab + thigh:wrist, data = bfatNormalized))
[1] -500.4447
> #compare to best linear model without interactions
> AIC(lm(pcfat ~ age + weight + neck + ab + hip + thigh +
+ forearm + wrist, data=bfatNormalized))
[1] -486.8114
```

This new "best" model is superior to the best linear model without interactions. Calculating the AIC_{weight} for only these two models and using AIC_{weight} as a selection criterion favors the interaction model.

Formula	AIC	AIC_{weight}
weight + height + neck + ab + thigh + wrist + weight:ab + weight:wrist + height:neck + height:ab + neck:ab + thigh:wrist	-500.4447	99.89
age + weight + neck + ab + hip + thigh + forearm + wrist	-486.8114	0.1

Table 6.3: bodyfat model comparison on AIC_{weight}

Backward elimination is a search heuristic. It cannot be guaranteed that the model above is the best. Using exhaustive search on the above variables with interactions produces 10 better models regarding *AIC* quality. The best looks as follows:

```
Call:
lm(formula = pcfat ~ weight + height + neck + ab + thigh + wrist +
    weight:ab + neck:ab + ab:wrist + thigh:wrist, data = bfatNormalized)

Residuals:
```

```
        Min         1Q      Median         3Q        Max
-0.218884 -0.053817 -0.004127   0.059628   0.213499


Coefficients:
              Estimate Std. Error t value Pr(>|t|)
(Intercept)    -0.7352     1.6677  -0.441  0.65971
weight          3.1049     1.2753   2.435  0.01563 *
height         -0.5839     0.2733  -2.136  0.03369 *
neck           -4.5739     2.1919  -2.087  0.03796 *
ab             -6.5839     2.9681  -2.218  0.02747 *
thigh           9.8513     3.1274   3.150  0.00184 **
wrist           2.0630     1.9569   1.054  0.29283
weight:ab      -5.3076     1.9050  -2.786  0.00576 **
neck:ab         6.9506     3.4929   1.990  0.04773 *
ab:wrist        7.8622     3.2143   2.446  0.01516 *
thigh:wrist   -11.3505     3.6351  -3.122  0.00201 **
---
Signif. codes:  0 '***' 0.001 '**' 0.01 '*' 0.05 '.' 0.1 ' ' 1

Residual standard error: 0.08708 on 241 degrees of freedom
Multiple R-squared: 0.7654,      Adjusted R-squared: 0.7557
F-statistic: 78.64 on 10 and 241 DF,  p-value: < 2.2e-16


> AIC(lm(formula = pcfat ~ weight + height + neck + ab + thigh +
+ wrist + weight:ab + neck:ab + ab:wrist + thigh:wrist, data =
+ bfatNormalized))
[1] -502.3158
```

Restricting the model sets to models with a better *AIC* than −500 is an option. The restricted[10] model set contains 28 models. These models cover only another small part of model space. Nonetheless these models lead to another idea what variables should be considered for the final model set. Using Mondrian the best coefficients can be chosen with a combination of *missing value plot* and the *p − value of the coefficients*

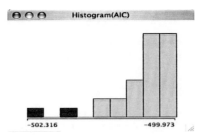

Figure 6.37: bodyfat, reduced variables: AIC selection of the best models including two-way interaction (Mondrian)

---

[10]on the above 6 variables

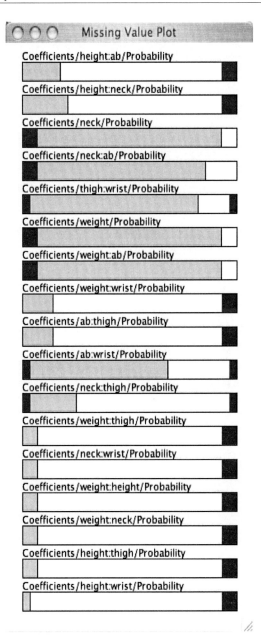

Figure 6.38: bodyfat, reduced variables: missing values plot to identify the coefficients (Mondrian)

6.5.2.2 An Alternative Approach To Model Selection

Given the huge model space, where only a few special models are to be considered for a candidate set other model attributes might be taken into account as well. As seen above many coefficient values do not contribute much with regard to coefficient factor weight. Using modern computer hardware[11] all 2^{91} combinations could be examined in the same way as presented above. But this kind of "brute force" exhaustive method is bound to provide poor results. Why should a factor combined of low-weight factors lead to a superior model ?

Nonetheless an important question is what factors have the most impact on the body weight. To fit better models in terms of quality more parameters are required. The additional interactions that have been found before in combination with all eligible variables might provide some superior models. The model search space contains 13 base variables and 15 interactions, that covers $2^{28} = 268435456$. As a threshold quality -500 is chosen and additionally a minimum coefficient weight of 0.2. This combination filters out globally bad models and ensures that each model with any near zero coefficient value is rejected. 50 models fulfill the above requirements.

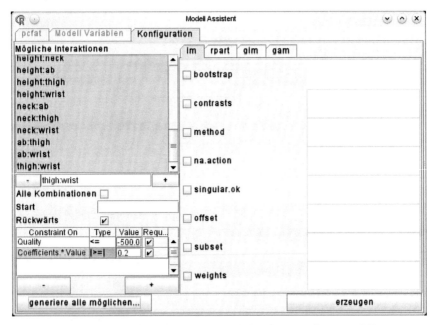

Figure 6.39: MORET GUI (Linux): creating models based on a combination of filter criteria

These models share a set of uncommon properties. All coefficients provide a minimal non zero weight and the overall model quality is good. To continue the variable selection process the

---

[11]not a single computer but a *really* large cluster. A modern home PC can fit about $10^8 - 10^9$ per day.

model quality is selected and inspected concurrently with a missing value plot. The missing value plot shows which variables are used in the model and which are omitted. Not all eligible interactions were selected in the course of this constrained model selection process. From the total of 28 only 22 are displayed. Overall 25 variables were selected, but each variable that is found in all models is omitted in a missing value plot. The parameters *age*, *ankle* and *knee* are not found in any of these models. The following set of figures (cf. table 6.39) is to be read from left to right. On the top part a histogram displays the selected model sets quality. In the terms of candidate set it is a best of the best models. Only few models were good enough. In combination with the missing value plot the variable selection frequency can be observed step by step. The first column left is one best model. All selected variables - plus *ab*, *thigh* and *wrist* that are found in all of these models - are marked in the missing value plot below. So the model is $ab + neck + thigh + wrist + height + weight + ab : neck + ab : weight + ab : wrist + thigh : wrist$.

Using a broader interval of models on the next column new model parameters are added. Step by step the top contenders, with respect to selection frequency, can be identified. Some variables do not show up in this best of the best candidate set, like *ab : height* or *height : thigh*.

Model averaging will hide details that are found on that level of step by step analysis. Parameters that are chosen rarely do not play a major role in such an averaged model. The question here is how many models should be chosen for a candidate set. A graphic method provides feedback on the variables found for any selected model set.

Another helpful plot that is regularly used concurrently is the parallel box-plot. This plot reveals when model parameters of both positive and negative sign are used. Instead of showing such a plot for each selected candidate set, a static box plot is used (cf. figure 6.40). Interactions are omitted in this plot because some compound variables (e.g *neck* or *height*) do already cover positive and negative values. Averaging might even end near zero in that situation - so it is one idea to split candidate sets and check which interaction leads to change of sign.

It is clear that model averaging and candidate set selection is no trivial task. This example shows the difficulties encountered to describe a rule for automatic model set supervision.

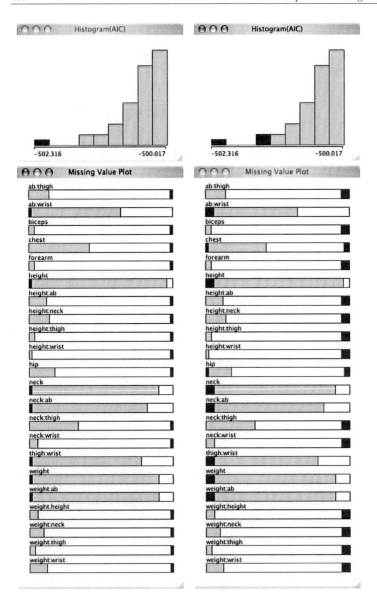

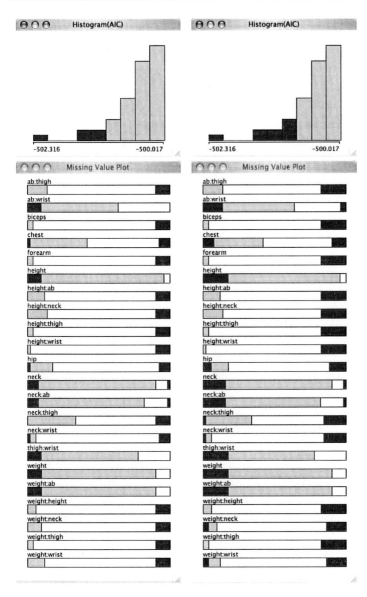

Figure 6.39: stepwise analysis of selected model parameters using Mondrians missing value plots

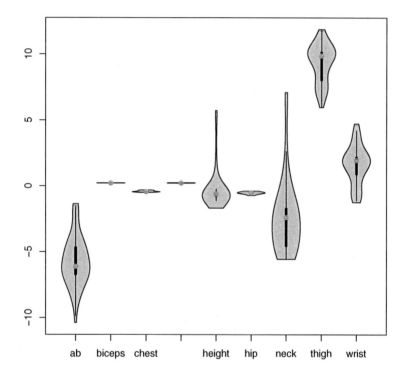

Figure 6.40: R violin plot: distribution of compound parameter weights

Some weights are purely positive or negative but some cover a wide range from negative to positive. Averaging will reduce these effects as shown before. Table 6.4 shows the parameter ranges, frequency and averaged effect.

Parameter	Frequency	Minimum	Maximum	Averaged	Averaged[12]
ab	50	-10.38	-1.35	-5.69	-5.67
biceps	2	0.20	0.23	0.01	0.01
chest	21	-0.50	-0.31	-0.17	-0.16
forearm	2	0.20	0.24	0.01	0.01
height	48	-1.68	5.73	-0.21	-0.23
hip	9	-0.71	-0.39	-0.09	-0.09
neck	45	-5.56	7.11	-2.07	-1.81
thigh	50	5.95	11.87	9.32	9.42
weight	45	-3.77	5.47	2.36	2.39
wrist	50	-1.24	4.73	1.62	1.74
ab:height	6	0.62	6.24	0.40	0.37
ab:neck	41	6.46	15.02	8.46	8.30
ab:thigh	7	-2.82	2.15	-0.08	-0.07
ab:weight	45	-6.84	-1.90	-4.64	-4.69
ab:wrist	32	1.02	9.37	4.81	5.08
height:neck	7	-13.80	-1.63	-1.12	-1.04
height:thigh	2	1.01	5.34	0.13	0.11
height:weight	4	0.21	5.12	0.12	0.10
height:wrist	1	0.61	0.61	0.01	0.01
neck:thigh	17	-11.61	-3.47	-3.01	-2.61
neck:weight	5	-7.43	-1.93	-0.51	-0.55
neck:wrist	3	-4.27	-1.72	-0.20	-0.21
thigh:weight	2	0.53	1.77	0.05	0.04
thigh:wrist	39	-13.56	-1.91	-8.24	-8.68
weight:wrist	5	0.68	8.59	0.82	0.89

Table 6.4: variable selection frequency and averaged parameter estimate

Parameter frequency directly affects averaged parameter estimates. In a set of 50 models any variable chosen only once or twice does not have much impact in terms of forecast weight. Interactive tools provide means to analyze the models in combination with the data. Without demonstrating these final steps the model space has been reduced from 91 variables - including interaction - to 25. Some interactions may be dropped, but the main point is that a next to unmanageable model space has been reduced to a number that can be worked on.

6.5.2.3 Averaged model properties

Model averaging is plausible when many good models exist and a single best model does not suffice. The origin of this insufficiency is any bias named earlier. But why should these models be aggregated ?

---

[12]weighted by AIC_{weight}

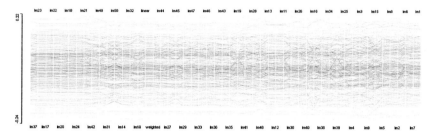

Figure 6.41: Residuals from single and averaged models as parallel coordinate plot

The figures 6.41 and 6.42 shows that the residual distribution varies even when models are very similar. Parallel lines emphasize that all cases are fit equally by all models. A close look shows parallel lines for the two models labelled weighted and linear. These refer to model ensembles, one with equally weighted residuals and one using Akaike weights. Other models like lm23 and lm17 fit most observations equally well but this frequency of parallel residuals is not commonly found among single models. Last, some observations along the borders should be paid a second look since no model is capable of providing a close forecast.

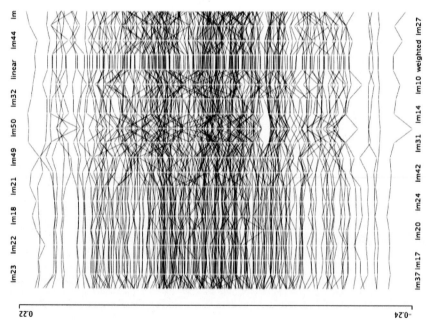

Figure 6.42: Zoomed part of the parallel coordinate plot from figure 6.41

One aspect is that the residual distribution might be skewed on some of the models. Averaging leads to a smoother distribution in our example (see figure 6.43). As a reference a gaussian distribution with an averaged standard deviation is included.

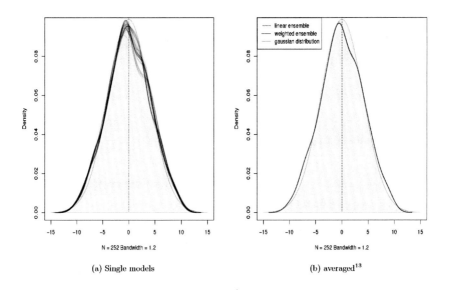

(a) Single models   (b) averaged[13]

Figure 6.43: Comparison of single model residual distribution versus ensembles

Accuracy of model ensembles

Figure 6.43 shows that the averaged residuals distribution is nicer that any single model distribution. But what benefit in terms of prediction accuracy is found in a model ensemble?

---

[13]The density of both ensembles is very close $\sqrt{(weighted - linear)^2} = 6.455569e - 05$. The plot is unable to separate both lines.

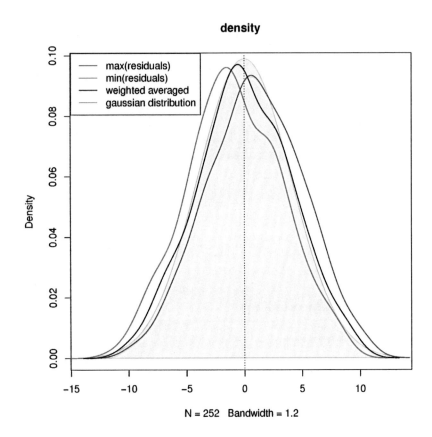

Figure 6.44: Extreme residual density in comparison to ensemble density

Figure 6.44 shows the minimal and maximal residuals as density plot in comparison with the averaged residuals. Naturally the averaged residuals are located between these borders. As a reference the gaussian distribution is included . An interesting question is, whether an ensemble is more accurate than the best model? The exactness of the model ensembles is not globally better than the best single model contained in the example. Nevertheless some observations may by incidence be fitted better by an ensemble than any single model is able to. Misusing ensembles for that purpose[14] is a very special kind of over-fitting. Real life data sets do seldom offer models that feature the required properties.

---

[14]i.e chose model ensembles in a way that only optimizes accuracy

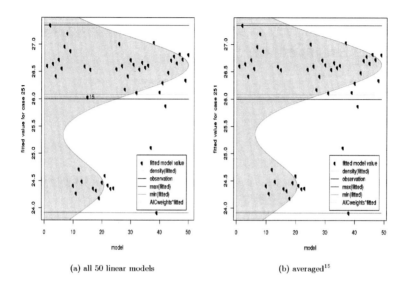

(a) all 50 linear models   (b) averaged[15]

Figure 6.45: model forecast precision at one observation

Forecasting from a model does not guaranteed to end up close to the actual observation. In the rare case when no model is really close and the forecasts balance around too high and too low the averaged forecast is superior to any single model forecast. Figure 6.45 is a real life situation where model 15 is in fact closer than the averaged forecast. The density of the residuals is displayed in transparent blue color.

```
> #averaged value
> as.double(bfat.bin%*%aicweights)
[1] 26.08171
> lm15$fitted[251]
[1] 26.03527
> pcfat[251]
[1] 26
```

If the ensemble does not contain model 15 or even a model that forecasts this observation lower the ensemble will be more accurate than any single model at this special observation. Optimizing an ensemble to predict a single observations too accurate is not a good idea. In fact it is a special case of over-fitting. Nonetheless this holds the potential for artificial examples that excel in prediction accuracy.

Best model versus model ensembles

Not all model ensembles beat the best model in terms of accuracy. But prediction accuracy is not the ultimate goal in statistic modeling. Preconditions must be met. Residuals that distribute widely in an unknown pattern are not conveniently fitted by linear models. Relying on a single model statistic for model selection is a disputable practice. Since ensembles have been chosen by some criteria the performance on that criterion is good, too. But in comparison to single best models they perform well over all. To emphasize this fact single models and ensembles are ranked on some criteria:

min	ranked on residual minimum (only negative residuals)
max	ranked on residual maximum (positive only)
median	ranked on absolute residual median
mean	ranked on absolute mean
RSS	residual sum of squares
AIC	averaged Akaike information criterion

Ensemble	min	max	median	mean	RSS	AIC
linear	26	19	20	26	4	21
weighted	25	22	19	27	5	16

Table 6.5: ensemble model criteria compared on ranks

It is no surprise to find many average ranks; the worst rank is 52 since 50 models and two ensembles are compared. In contrast those models that excel in one criterion perform poorly in the others. Figure 6.46 displays bright green for best ranks and bright red for the worst. Dark color represents mediocre ranks.

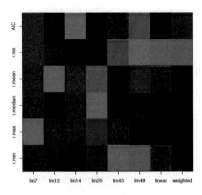

Figure 6.46: Model ensemble quality statistics compared to best models

Conclusions

The choice should be made in favor of model ensembles when no clear best model suffices. Although model ensembles can not outperform any single model in every respect, the averaged model is better balanced and performs well all-round.

What can be learned from such an arbitrary chosen model set ?

☞ constrained model generation greatly reduces the number of valid models

+ reduces model space

+ spurious effects are reduced

- extra validation required with respect to chosen bias

☞ a data set might reveal distinct disjunct[16] candidate model sets

Filters provide an effective mean to constrain model space. In combination with an interactive graphical tool subsets of models can be analyzed effectively. Interactive selection on the coefficient influence (see figure 6.36 on p. 125) leverages variable selection. Intervals for forecasting can be determined. Interactions and implied structural changes become evident at a glance.

Additional possibilities like the comparison with cleaned up data emphasizes coefficient selection, see figure 6.39 for an example. Other model selection strategies can be used with MORET. Using a hard filter criterion introduces a bias in model selection. Since good versus bad models on a given quality criterion are favorable this step can be taken earlier or later. Even special model bias can be achieved using filters. E.g only models with a given minimal coefficient influence (§6.3.2 on pages 99) might be considered, or all coefficient p-values can be a subject to constraint.

6.6 Summary

Implementing the useful features as described in the design section (§5.3, p. 62) MORET provides a tool to manage models and related data. Lots of R model types can be customized and added without effort. The collected data can be assessed in small quantities that are interactively filterable and sortable. The *model overview table* provides global model selection while deeper insight is gained using the *model explorer*. Alternatively a selected set of model data can be exported in different computer readable formats for further processing. The administered model data can be used for many purposes. Two examples are provided:

✔ Once a spreadsheet program was fed directly from the model explorer tool.

✔ On the other occasion a model set was transformed into a special format for further analysis with an interactive graphic analysis tool (Mondrian Theus (2002b)).

Additional useful extensions like the database import and export facilities were only mentioned briefly. This features support storing and recovering the model data base to/from a neutral format. That provides an interface for model sharing without the necessity for central data management. MORET can be used as a central tool collecting and administrating models. Everybody needing to administer sets of models can find valuable support using MORET.

---

[16]Consider to split candidate sets when different signs for compound variables are encountered.

Chapter 7

Conclusion

7.1 Summary

Statistical models serve many purposes. Analyzing the interaction of variables leads to a better understanding of our world. Separating noise and structure allows extrapolation. This can be used to forecast future development. A wide range of applications exists. Yet most statisticians rely on a single model to draw conclusions. Multi model administration has been too laborious. This work deals with the problem of model administration and multi model inference. In the introductory chapter 1 statistical models are introduced. A few linear models fit on the hills data set serve as an example to compare models on different information levels. Additionally the concepts of model selection bias and model uncertainty are introduced. These data and method related problems are the main reasons to employ more than a single best model for inference.

Some theory to compare models is presented in chapter 2. The Kullback-Leibler Information is introduced as a general framework for comparing models. Comparison on the highest level of information is common practice. An extensive list of information criteria is presented.

In the following chapter 3 the topic of model selection is given detailed consideration. Model validation is discussed and ensemble modeling introduced. Modern modeling techniques like Boosting or ensemble BMA are discussed along the way.

After that chapter 4 presents interactive model analysis as the tool of choice to analyze model data. Chapter 5 deals with a US presidential election data set. No single dominant model needs to be found for that application as the question is to find out if one binary variable has a significant impact on the results. Models are fitted manually to motivate the necessity for a model administration software. The theoretical requirements for this software are presented and useful tools discussed. In chapter 6 the analysis of the presidential data set continues, this time with software support. Different points of view are taken on the selection of a candidate set. A second data set is also considered to compare sets of models. This time the *bodyfat* data set is used. The full data set is compared with a cleaned up subset of the original data.

7.2 Achievements

The purpose of this work was to design and implement a software to manage a large amount of models. This aim has been met by the software named MORET. Furthermore many useful tools have been developed. Not all of the tools can be presented in full detail within the scope of this work. Still the current overview shows some interesting advances in model administration and ensemble method management.

Model Configuration, Storage And Management

While the algorithms are of minor interest for a statistician technical explanation is necessary. The details are presented to help working effectively with this new tool. R provided a super-abundance of model packages. More packages exist than distinguishable model classes. MORET provides means to compare any type of model. Prior configuration of the R model objects is a requirement though.

The model database, that provides the data for candidate set selection is filled by models generated with the special MORET GUI. Any type of R command that supports a complexity and quality measure may be processed as a model. Analyzing and managing these stored models requires tools to determine a list of models, either by model attributes or by global criteria.

Model Fitting Under Constraints

Since model space is large and to avoid known bias by stopping rule or partial model space scan, MORET is able to scan a predefined subset of the model space and store exclusively models that are subject to an arbitrary number of constraints. A practical application is to filter out all models that feature any insignificant coefficient; insignificant means that a threshold p-value is required.

Model Ensemble Selection By Interactive Means

Model selection, especially for ensemble methods, should not rely on high level statistics alone. Variable dependent statistics like coefficient p-values should be taken into account, too. As shown with two exemplary data sets it is unreasonable to manage that amount of information manually. The analysis and comparison of high and intermediate level statistics should be performed interactively. This new tool even supports the interactive comparison of derived statistics, e.g AIC_{weight}.

MORET provides the technical tools for model and model ensemble administration. Model variable selection using model averaging techniques needs extra care. Interactions that change the sign of compound variables skew averaging effects. Thus the better solution is to supervise candidate sets manually.

7.3 Next Steps

MORET offers new tools and this work covers some applications of the tools. The methods created should be employed by practitioners to refine multi model techniques. Especially averaging techniques should be validated and refined. What is to be done when the candidate set includes interaction? Is there a useful way to employ averaging in this case or does this lead to disjunct candidate sets? Theory about averaging on different sub-samples is scarce. Apart from bootstrap applications this combination has not been explored. Multi model inference is used to reduce risks - estimators for these risks should be elaborated.

Much work and research could be done in the field of multi model inference. I hope this work provides good tools to facilitate exploration in the area.

Appendix

A.1 R Model Packages

This section is devoted to present a selection of extension packages to the base R software (R Development Core Team, 2006). Some packages provide specialized models, others employ various ensemble techniques. An overview of all R extensions can be found in the internet at http://cran.r-project.org/ using the link to **packages**.

package	description
BMA	Bayesian Model Averaging
caret	Classification and Regression Training
ensembleBMA	Ensembles and Bayesian Model Averaging
maptree	Mapping, pruning, and graphing tree models
mboost	provides the boosting procedure for popular models
meifly	Interactive model exploration using GGobi
randomForest	Breiman and Cutler's random forests for classification and regression
TIMP	a problem solving environment for fitting superposition models

Table A.1: A selection of model related R packages

A.1.1 Examples: Bayesian Model Analysis

Revisiting the "hills" data set the R-package *BMA* provides methods to automatically create a candidate model set. The next short example shows the usage of the function MC3.REG that performs simultaneous variable selection using Markov chain Monte Carlo model composition (MC3). Other supported functions involve BMA variable selection for linear regression, generalized linear models or survival models.
Just for reference, the variables and indices:

1	2	3
dist	climb	time

```
> b<-c(6,7,18,33)
#use "recommended outliers instead of computed"
# out.ltsreg(hills[,-3], hills[,3], 2)
> hills.mc3<-MC3.REG(hills[,3], hills[,-3], num.its=20000, c(FALSE,
TRUE), rep(TRUE,length(b)), b, PI = .1, K = 7, nu = .2, lambda =
.1684, phi = 9.2)
> summary(hills.mc3)


Call:
MC3.REG(all.y = hills[, 3], all.x = hills[, -3], num.its = 20000,
M0.var = c(FALSE, TRUE), M0.out = rep(TRUE, length(b)),
outs.list = b,  PI = 0.1, K = 7, nu = 0.2, lambda = 0.1684, phi = 9.2)

Model parameters: PI = 0.1 K = 7 nu = 0.2 lambda = 0.1684 phi = 9.2

  20  models were selected
 Best  5  models (cumulative posterior probability =  1 ):
```

	prob	model 1	model 2	model 3	model 4	model 5
variables						
dist	1	x	x	x	x	x
climb	1	x	x	x	x	x
outliers						
6	0.07996	.	x	.	x	.
7	0.99997	x	x	x	x	.
18	1.00000	x	x	x	x	x
33	0.93337	x	x	.	.	.
post prob		0.8553773	0.0779860	0.0646268	0.0019762	0.0000279

Since the best 5 models use both *dist* and *climb* this would suggest to use both variables for inference.

The MC3.REG will also identify outliers for each model, for the settings used the outliers are included in the analysis, but it is possible to exclude outliers automatically using the outliers=FALSE switch:

```
> hills.mc3.no<-MC3.REG(hills[,3], hills[,-3], num.its=20000, c(FALSE,
TRUE), PI = .1, K = 7, nu = .2, lambda = .1684, phi = 9.2, outliers=
FALSE)
> summary(hills.mc3-no)


Call:
MC3.REG(all.y = hills[, 3], all.x = hills[, -3], num.its = 20000,
M0.var = c(FALSE, TRUE), outliers = FALSE, PI = 0.1, K = 7,
nu = 0.2, lambda = 0.1684, phi = 9.2)
```

```
Model parameters: PI = 0.1 K = 7 nu = 0.2 lambda = 0.1684 phi = 9.2

  4  models were selected
 Best  4  models (cumulative posterior probability =  1 ):

            prob    model 1    model 2    model 3    model 4
variables
  dist     1.0000  x          x          .          .
  climb    0.9995  x          .          x          .

post prob          9.995e-01  5.304e-04  2.947e-10  1.690e-16
```

Utilizing the *bicreg* functions yields (in that case) the same result as the simple linear model would have.

```
bicreg(x=hills[,-3],y=hills[,3])

Call:
bicreg(x = hills[, -3], y = hills[, 3])


 Posterior probabilities(%):
 dist climb
  100   100

 Coefficient posterior expected values:
(Intercept)         dist          climb
   -8.99204      6.21796        0.01105
> lm(time~dist+climb)

Call:
lm(formula = time ~ dist + climb)

Coefficients:
(Intercept)         dist          climb
   -8.99204      6.21796        0.01105
```

In any case the Bayesian Model Averaging BMA package selects the same model without interaction.

Technical Details

Many technical details that should be subject to further explanation remain hidden deeply inside MORET. This chapter reveals how the model values are derived from *R*. The algorithm used to transform the R object as provided by Rserve (Urbanek, 2003) into a form that is suitable for database is presented. Since the R user environment, called workspace, may contain many objects at once, alternative ways exist when this algorithm may be executed. The best practice is discussed at the end of this chapter.

B.1 Extracting data from R

Note: This section is very specific to the software R.
MORET needs all relevant data to store inside a relational database. Binary objects cannot be handled fast enough to compare different aspects of similar models like *lm* and *glm*. In this context similar means that the model summaries will be very much the same if the same formula is used.:

```
> usa.lm1<-lm(fl_b04~fl_d96+fl_b00+fl_his+fl_inc+fl_eto)
> usa.glm1<-glm(fl_b04~fl_d96+fl_b00+fl_his+fl_inc+fl_eto)
> summary(usa.lm1)

Call:
lm(formula = fl_b04 ~ fl_d96 + fl_b00 + fl_his + fl_inc + fl_eto)

Residuals:
      Min         1Q     Median         3Q        Max
-0.0570926 -0.0135550  0.0008711  0.0148481  0.0555068

Coefficients:
              Estimate Std. Error t value Pr(>|t|)
(Intercept)  2.624e-02  2.602e-02   1.009   0.3172
fl_d96      -2.275e-01  1.226e-01  -1.855   0.0684 .
```

```
fl_b00       1.299e+00  1.029e-01  12.632   <2e-16 ***
fl_his      -5.193e-02  2.981e-02  -1.742   0.0865 .
fl_inc      -1.080e-06  7.087e-07  -1.524   0.1326
fl_eto       1.095e-03  7.537e-03   0.145   0.8849
---
Signif. codes:  0 '***' 0.001 '**' 0.01 '*' 0.05 '.' 0.1 ' ' 1

Residual standard error: 0.02319 on 61 degrees of freedom
Multiple R-Squared: 0.9572,      Adjusted R-squared: 0.9537
F-statistic: 273.1 on 5 and 61 DF,  p-value: < 2.2e-16


> summary(usa.glm1)

Call:
glm(formula = fl_b04 ~ fl_d96 + fl_b00 + fl_his + fl_inc + fl_eto)

Deviance Residuals:
       Min         1Q      Median         3Q         Max
-0.0570926  -0.0135550   0.0008711   0.0148481   0.0555068

Coefficients:
              Estimate Std. Error t value Pr(>|t|)
(Intercept)  2.624e-02  2.602e-02   1.009   0.3172
fl_d96      -2.275e-01  1.226e-01  -1.855   0.0684 .
fl_b00       1.299e+00  1.029e-01  12.632   <2e-16 ***
fl_his      -5.193e-02  2.981e-02  -1.742   0.0865 .
fl_inc      -1.080e-06  7.087e-07  -1.524   0.1326
fl_eto       1.095e-03  7.537e-03   0.145   0.8849
---
Signif. codes:  0 '***' 0.001 '**' 0.01 '*' 0.05 '.' 0.1 ' ' 1

(Dispersion parameter for gaussian family taken to be 0.0005378819)

    Null deviance: 0.767417  on 66  degrees of freedom
Residual deviance: 0.032811  on 61  degrees of freedom
AIC: -306.52

Number of Fisher Scoring iterations: 2
```

The coefficients are in fact identical, still *glm* uses other global quality measures. *glm* uses *AIC* per default and *lm* R^2. In this respect the *summary*-objects differ, though their forecast and shape is identical. Special packages support the extraction of *AIC* for linear models too, so by employing careful preparation identical values are obtained.

When it comes to extract quality measures from *R*-objects the first step to do is to decide when the data shall be derived from the *R*-object and written to the database.

B.1.1 When To Extract Data

Since the data has to be calculated in R first, nothing can be done until R is finished with the internal calculation. This is technically true but does not cover the whole truth. The MORET-GUI handles the commands that are really sent to R. So when MORET recognizes a *model-command* by a known pattern internal things may happen first.

By default MORET knows about

- lm - linear models

- glm - generalized linear models

- gam - generalized additive models

- rpart - classification and regression tress

And in very little time, following the tutorial from the web-page *loess* can be introduced to MORET. Now if any of these commands are supplied, the way they are used in R, MORET will not simply delegate the calculation through Rserve to R. A predefined algorithm will check some integrity constraints and only save the created model if no error has occurred.

Algorithm 4: Model Extraction Algorithm

Input: :

$\mathcal{I}$ user input string **Output**: $\mathcal{M}$ model data

parse user input;

if *user input command is known in the model database* **then**
>   stop computation;
>   restore binary model;
>   output "warning, known model";

else
>   execute computation via Rserve in R;
>   **if** *error during calculation* **then**
>>   notify user of error cause;
>>   stop computation;
>   **else**
>>   extract data from R-object;
>>   save binary file to file-system and database;
>>   save extracted data object to relational database;
>   **end**

end

Implementing this algorithm, though it looks very easy, proves technically very demanding. The first reason is that not everything that must be regarded as a calculation error will return an error-type via Rserve. Secondly R uses few generic data-types, but model summaries (or objects) are not always created in the same manner for "equal" structures and different model classes. Equal, in this context, refers to the presentation. Because of this structural non-uniformity additional measures need to be taken. E.g. working around the standard summary by defining an adjusted data container.

Data extraction will ultimately be handled via Rserve, but to find data there are at least two ways:

- use the generic data types from R through Rserve and map the resulting types into a tree like structure

- define specific *R*-commands to extract parts of the data and create a special mapping procedure for each special command. From these mappings collect all data and proceed as usual.

The first way is preferable since the user just needs to define the mappings from *R* to Rserve via configuration. The second option involves programming.

1. first enter an *R*-command into the configurator and check if a valid model command is automatically created in the line "Example". The Description is optional.
   Since the summary command is not always a sufficient data source a *custom command* will substitute the usage of `summary({model})`
   The above defines the data source within *R*, if the resulting tree should be pre-modified and this be done by adding an `xslt-template` to restructure the result.

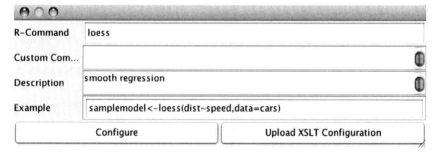

Figure B.1: MORET GUI: start a simple default model configuration (for another example see figure 6.2 on page 93.

2. from the resulting model summary object tree (left hand side) drag and drop a node for overall quality and for the complexity into the input fields "DF" and "ResidualDeviance".

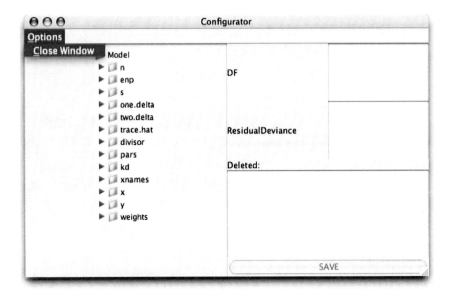

Figure B.2: MORET GUI: configure the model by drag and drop (compare 6.3, p. 94)

3. Drag and drop all nodes that should be ignored by the database into the "Deleted"-Area. The tree structure is modifiable via drag and drop, too. Save the configuration and create models - if an error occurred during this configuration step the configuration can be deleted and reconfigured from scratch.

More considerations about the time, *when* the model data is to be extracted from R must be taken into account. A possible approach is to check the workspace for model objects: synchronously or asynchronously search R's workspace for new objects. If a model object is identified use algorithm 4 to store the data persistently.

This idea could not be implemented efficiently since the models are created in a single instance of this workspace. Synchronously the user is forced to wait until the checking has been done after each command entered. Worse, the delay will increases with the number of objects in the workspace. Even asynchronously the application slows down with workspace size. The reason is the single workspace instance (and for windows based platforms single threaded connection). This idea has not been fully implemented because of the slow performance.

B.1.2 Mapping Data From R

In most cases Rserve is able to extract all of the data that is contained in the summary object. In rare, obscure cases some important data gets mysteriously lost. In these cases there is no way around writing a custom mapping procedure. For a language reference on R either use R's integrated help, also directly accessible from MORET. Before implementing this special mapping it is highly recommended to create block-wise commands to send through Rserve, as a list of single extraction command is computationally very expensive. Useful R-commands are *names*(∗), *rownames*(∗), *typeof*(∗), *as.list*(∗),*as.vector*(∗),*as.double*(∗), ... Other type conversion e.g. from named array to unnamed array may be used, too.

Four special mapping algorithms are presently found in the sources to serve as example templates. (see p. 150) The most simple mapping is found for the linear model. To create a mapping, first examine the model-object in R to see what should be stored permanently. Take a look as well at the summary report. From these R-objects - mostly list structures - the interesting values are picked for safe-keeping. In the case of a linear model the coefficients are picked as specific structure as well as R^2 and R^2_{adj} (cf. figure 5.8, p. 64). Degrees of freedom are a best choice for model complexity. *AIC* is a versatile and common quality statistic that is available for many model types and thus worth using as a default quality measure. *Deviance* is an alternative quality measure that is found in linear models and generalized linear models and comparable on an equal scale

The technical mapping can be derived from the source, see the package **de.augsburg.uni.-rosuda.model.type.lm**; the class that pulls the values from R is **RLinearModelExtender**. **LM** is the controlling class that is registered as a trigger command in the database. The procedure used conforms to algorithm 4 (p. 150). Another noteworthy part is how the this process is triggered:

B.1.3 UML sequence diagram: intercepting user commands

Figure B.3 documents the user interaction with R (via Rserve) and the database. The graphical user environment allows textual input of R commands. Each command is passed to a pre parser. This parser decides if this is a special model command that should be stored in the database or if this is an unknown command. Unknown commands are just passed to Rserve and the response will be returned to the user interface.

In the other case, when a user enters a model command like *lm* the parser starts a special process to extract all model values and store the data into the database. Algorithm 4 (p. 150) emphasizes the effect that a model is defined by its data and model command. MORET will not re-calculate a model if the same model has been created before. This aspect is left out of the diagram to keep focus on the important parts. The interception (figure B.3 on p. 154) only emphasizes the moment when the special process is executed. The data extraction process is by far the most complex part and many solutions have been tried out during the creation of the software MORET. The software R does not restrict future development to support older modules, at least not in every respect. Since this is not granted, configuration is regularly required. Minimally after each new major version of R. Most changes are usually not very complicated like moving a degrees of freedom measure inside the summary report. Nevertheless MORET enforces a stable interface to guarantee comparability over models of the same structure (and some times even across classes).

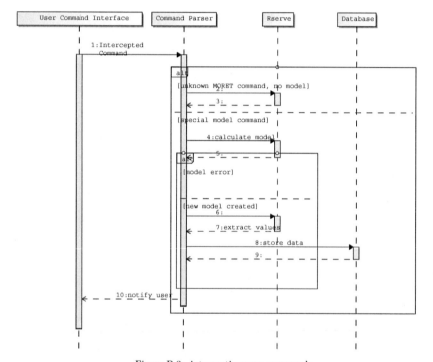

Figure B.3: intercepting user commands

It would be a nice addition to *R* if classes guarantee backward compatibility. Changing *R* objects for non-public research modules is okay but released packages rely on other packages. Changing object structure will result in errors on all dependent software.

B.2 R source code

This section covers *R* sources used in MORET.

B.2.1 Create Bootstrap data.frame From data.frame

```
 sample.dataframe<-function(dataset)
{
    rows<-dim(dataset)[1]
    cols<-dim(dataset)[2]
    #matrix for bootstrapped data from dataset
    mydataset<-matrix(nrow=rows,ncol=cols)
    rownames(mydataset)<-rownames(dataset)
    colnames(mydataset)<-colnames(dataset)
    sampleidx<-sample(1:dim(dataset)[1],replace=TRUE);
    #row index by bootstrap for new sampled bootstrap dataset
        #copy data into bootstrap sample matrix
        for (r in 1:rows)     {
        for (c in 1:cols){
            mydataset[r,c]<-dataset[sampleidx[r],c]
        }
    }
    data.frame(mydataset)
}
```

B.2.2 Extract R^2_{adj} For Linear Model

```
lmmodelquality<-function(dataset,formulae)
{
    v<-rep(0,times = length(formulae))
    for (i in 1:length(formulae))
    {
        lmx <- lm(formula=formulae[[i]],data=dataset)
        v[i]<-summary(lmx)$adj.r.squared
    }
    v
 }
```

B.2.3 Create Rank Matrix For Linear Models

```
lmbumpmodels<-function(iter=10,formulae,dataset){
    columns<-length(formulae)
    mat<-matrix(nrow=iter,ncol=columns)#result matrix
    colnames(mat)<- as.character(formulae)
     for (row in 1:iter){
        bump<-lmmodelquality(dataset=sample.dataframe(dataset),
+ formulae=formulae)
         for (col in 1:columns)
        {
            result<-rank(bump)
            mat[row,col]<-result[col]
        }
    }
    mat # return matrix
 }
```

B.3 regular expressions

Regular expressions represent a very powerful way to describe a series of characters. Expressions can be used to match any type of character sequence. For example large text documents can be matched or lots of small text fragments filtered. MORET uses regular expression, commonly called *regex*, to filter from the model table (see e.g. figure 6.12 on p. 105). A *regex* is equally powerful as a *non-deterministic finite automaton* see Brüggemann-Klein (1993). Each node on the finite automaton represents a character in the sense that a *regex* can be understood. If you want to create an automaton that validates MORET you could use an automaton that looks like :

Figure B.4: finite state automaton

The automaton starts with the first character M, has only one edge (transition) to follow to O, and from O to R, R to E and finally from E to T. So this automaton matches exactly one word: MORET.

Using lettering and language everybody is capable of reading and understanding such simple automata. The regular expression language has developed as an easy, short alternative representation of such graphs. Since the same alphabet is used for the characters that represent nodes and the ones representing edges some restrictions arise which characters can be used "as read" and what characters need a special treatment so that a computer can discern if the character represents an edge or a node.

Sample B.4 shows that **MORET** can be a regular expression and will exactly match *MORET* - capital letters match capital letters. Since graphs can become very large it makes sense to allow subgraphs, grouping parts of a graph for better readability and more. For grouping brackets are used ( ) so brackets cannot be used directly for nodes. Alternative representations of **MORET** are **(MORET)** or other arbitrary sets of groupings like **(MORE)(T)**.

This grouping will make sense if a (sub-)graph, that is used a number of times, needs to be contained in the tested string to match the expression.

Figure B.5: MORE(T)*

Figure (B.5) matches e.g MORE, MORET, MORETT, MORETTT ...

B.3.1 Quantifiers

$\{n\}$ represents an absolute count of n times

$\{n,\}$ represents an absolute count of at least n times

$\{n,m\}$ represents an absolute count of $[n,m]$ times

? represents $0\ldots1$ times

* represents $0\ldots\infty$ times

+ represents $1\ldots\infty$ times

Table B.1: regex quantifiers

Note: All quantifiers from B.1 are called **greedy**, it is possible to postfix *?* or *+*. If *?* is postfixed the quantifiers are called **reluctant**, and in the *+* case **possessive**.

Connecting (sub-)graphs via the cardinality as above will provides quite a lot of variations. Nevertheless on many occasions it is very useful to connect (sub-)graphs as *alternatives*. The character | can be used for that purpose.

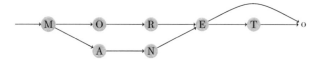

Figure B.6: M((OR)—(AN))ET?

Graph B.6 will match MORE, MORET, MANET and MANE.

B.3.2 character classes

Now if only literals (characters) were matched *regex* would be very long and matching words would be very tedious. For convenience expressions can also match predefined classes or subsets of the alphabet. The syntax for explicit matches uses brackets [], so $[a-z]$ can match any lower case character, $[A-Z]$ capitals and $[0-9]$ numbers. These can be combined to $[a-zA-Z]$. The ^ literal represents the complement (everything not matching the following expression) and can be combined with any other character class, like `[a-z&&[^cd]]` meaning each lower case character except c,d. `[^ot]` matches everything without o or t. Some special classes are predefined:

[H] . is a placeholder for any literal
\d   is used for any digit $\Leftrightarrow$ [0-9]
\D   is the complement of \d $\Leftrightarrow$ ^[0-9]

\s	short cuts whitespace characters \f, \n, \r, \t
\S	is the complement of \s so it matches any non-whitespace character
\w	is used for any "word" literal $\Leftrightarrow$ [a-zA-Z_0-9]
\W	is the complement of \w, any "special" character.

Table B.2: Predefined Character Classes

B.3.3 special characters

As proposed at the beginning the short concise writing of regular expressions introduces the problem that special characters that will for the result can not be used. Therefore a regular expression such a character \ + or some kind of bracket **must** be escaped by prefixing \ before the special character. \(matches the left round bracket (.

. arbitrary literal

? quantifier [0-1]

+ quantifier [1-∞]

* quantifier [0-∞]

& && is used as intersection

(,) grouping *capturing group*

[,] character class

{,} quantifier

| OR combination of literals or capturing groups

^ complement character or used for beginning of a line

$ end of a line

Table B.3: Special Characters

\f whitespace: form-feed

\n whitespace: line-feed, newline

\r whitespace: carriage-return

\t whitespace: tab

\On octal value for n (octal means $n \in [0 \dots 7]$)

\Onn octal value for nn

\Omnn octal value for mnn where $m \in [0 \dots 3]$

\xnn the ASCII character represented by the hexadecimal nn value (hexadecimal numbers are commonly found and written as [0..9a-f])

\unnnn unicode-representation of a the hexadecimal value $nnnn$

\e escape character \u001b

\a alert character \u0007

\cx control character x

Table B.4: Combined Special Characters

B.4 is not exhaustive but leaves out boundary characters $\backslash + x \in \{b, z, A, B, G, Z\}$ long quotation \Q \E and POSIX syntax $\backslash p\{x\}$. Since the *regex* interpreter used by MORET does not support all of these, this is just mentioned here. Regular expressions cover a wide range of applications that go beyond the scope of this work.

B.3.4 Using regex For MORET Table filtering

The similarity of *regex* and a graph (finite automaton) have been shown though not any way to construct the regular expression from the graph. Still the features of regular expressions have been presented by some simple examples. The internet contains many valuable resources on the topic and to name a single book Friedl (2006) can serve as a reference.

Regular expressions will be used in MORET to filter tables, the most common usage will be to find a substring or set of substrings (usually coefficients). So if you want to filter all models that at least contain "etouch" you can use the regular expression:

```
^(.*?etouch).*?$
```

If more than one substring must be found we need a *special construct*. This has not been introduced earlier. The syntax *(?=X)* matches X, via "zero-width positive lookahead". This allows that

```
^(?=.*?etouch)(?=.*?income).*?$
```

or

```
^(?=.*?income)(?=.*?etouch).*?$
```

can both match strings starting with

```
lm(b_change~etouch+income
```

without the need to pay attention to the order of the arguments. The special construct will only match if both arguments are encountered. It should be clear how an expression must be constructed to match three or more substrings. This syntax may seem weird but these examples illustrate how a powerful filter can be written with little effort. Constructing your own *regex* can be tedious but when the trick is found the results are very satisfying and moreover the filtering is quite fast.

B.3.5 Summary

Getting accustomed to regular expressions pays off since it supports powerful filters with few literals. When the syntax and characteristics are learned of how to construct such a *regex* (often it is a good idea to create a graph first because the overview is better) it is a very convenient way to achieve exact sub-selection with little effort. MORET manages incredible amount of models that may result in long, confusing lists - and these lists need to be filtered efficiently. Concise *regex* work as efficient filters. A more user friendly, yet effective alternative tool is hard to provide.
One final special *regex* is useful too, the (?!X) covers non-existence hence

```
^(?!.*?income)(?=.*?etouch).*?$
```

can find all models containing *etouch* but not *income*.

B.4 UML

UML is short for "Unified Modeling Language" and has been developed from about 1990. It unifies modeling for software and software systems. As everything related with software UML also comes in different versions. The actual 2.1 version covers many new ideas that were not found in the initial drafts. Many books cover aspects about UML but for this work only a superficial understanding is required. Knowledge about object oriented software comes in handy, though only three different diagram types can be found in this work:

1. class diagram

2. component diagram

3. sequence diagram

From object oriented software a few entities have to be explained shortly for easier understanding:

B.4.1 Nomenclature

B.4.1.1 Class

A class is a central concept of object–oriented software. As in the normal–world language class is used for any abstract thing that can be described by its specific behavior and its attributes. It is not compulsory that a class features attributes and special behavior but it is convenient to start with attributes and behavior to get in touch with the concept.

For instance an automobile is a standard example for a class. Each object that is from the class *automobile* features a finite number of axes, tires and seats. These attributes, i.e *number of axes* is described as numerical integer value. Each automobile has been build to move somewhere so it needs behavior like turn to the left, turn to the right, accelerate or break. These behavior will be presented by methods what the object should do. So a method for *turn* that will get an parameter for the direction and method(s) for the forward movement would be typical representatives.

A class is denoted by box. The top compartment contains the unique class name, the lower department contains the attributes and the lowest compartment the methods.

Note: Attributes and method all must provide *types* and *visibility*, these have special implications and are not of any relevance to understand the presented diagrams. For completeness sake, and not to be confused by the graphical representation, the first literal before the attribute or method represents the visibility:

- private

protected

~ package

+ public

Visibility limits the usage of the attribute or method from another related class.
The type gives information about "what is it", number, floating point number .. etc. Methods
must show a return methods denoted behind the closing bracket and the ": void" denotes that
no value is returned. Inside the methods brackets so called parameters are denoted that must
be provided to use the method, like the angle of how much the steering wheel is turned left.

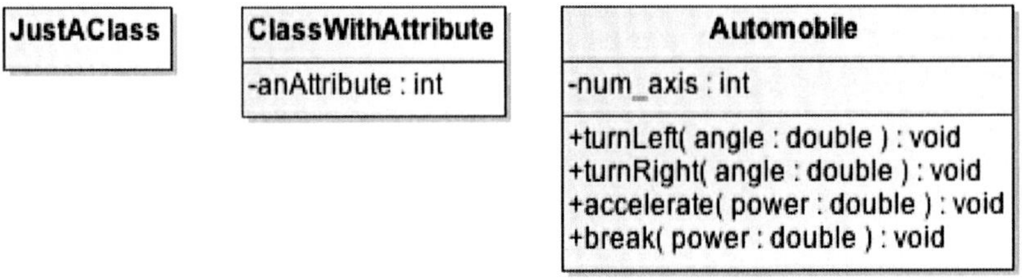

JustAClass

ClassWithAttribute
-anAttribute : int

Automobile
-num_axis : int
+turnLeft(angle : double) : void +turnRight(angle : double) : void +accelerate(power : double) : void +break(power : double) : void

Figure B.7: UML: graphical representation of a class

B.4.1.2 Relations

Another interesting fact about classes is that classes may relate to other classes. So a automobile
can be a container for passengers or cargo or the street net can be interpreted as a kind of habitat
for automobiles. There are different kind of class relations:

Inheritance   A sub-class that inherits from a super-class has access to the classes attributes and
behavior. Additionally the subclass may define own attributes or override super-class behavior.
Assume that the automobile needs to turn 90 degrees to the right. Most cars will only turn
the front wheels to the right but there are special cars that use four-wheel steering or rearwheel
steering. To achieve the same effect the internal implementation for "turn" with respect to
steering will differ.
Subclasses are drawn as a line with a triangle arrow head that points from the subclass to the
superclass. In figure B.8 A is the superclass of B.

Association   is the general term how classes relate to other classes. Associations have a specific
multiplicity ranging from zero to any arbitrary integer number. The association ends are called
role - the role each member of the association plays.
Relations are represented by lines between classes, if there is a relation between more than two
classes a rhombus is drawn at the intersection of the lines. If the classes need to be of a special
cardinality this required cardinality is written at the end of the arrow. 0..sth. tells that this
attribute is optional, 1 mandatory.
Note: At the end of each line there can be a small arrow → that indicates what class may
navigate that association.

Aggregation A common association is the aggregation - the belongs-to relationship. One class aggregates another class but the classes are not dependent on each other. For an automobile passengers would be a fit example. A passenger plays the role of in-the-automobile-sitter the car the role passenger-container. Both, the passenger as well as the car are independent classes so if the passenger leaves the car the car is (usually) not destroyed as well the passenger survives the getting out. Survival in the object oriented sense means that the object "lives" as long as it is references from somewhere. Any class needs to have some association to the other class. If a model is deleted from the database all of its data must be deleted. Even dependent data like attachments since these depend on the model. The script that was used to create the model will be deleted but not the data set from which the model had been built. Would we delete the data set all models and dependent data would be deleted in the same action since the model cannot exist without the provided data from the data set.

Composition This is the more stringent form of the belongs-to relationship. The compositions is a dependent belongs-to relationship. One class cannot exist without the other class. A car without a driver is still a car but a car without an engine is a trailer. So to construct a car of different class parts an engine of some kind is a compulsory attribute (among others). For a statistic model some data are required to create the model.

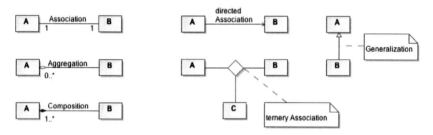

Figure B.8: UML: graphical representation of a class-relation

B.4.1.3 Interfaces And Abstract Classes

Interfaces and *abstract classes* define only behavior for a class. For the above automobile example "Movable" might be an interface of the class automobile and each class that implements such an interface is forced to provide all behavior defined in the interface. Abstract classes are much like interfaces but also offer attributes that can be made accessible by sub-classes. The point is that the behavioral part of classes can be defined on its own right and this behavioral part alone allows many algorithms to work since the algorithms can rely on the behavior. So as one example it is possible to write a sort algorithm that can sort any object that is comparable. Using behavioral structure so called "design patterns" offer solutions for many kind of problems that need not be addressed in special way but can be solved on an abstract level.

An Interface is denoted by a circle on the right side of the top compartment. An abstract class is denoted by printing the class name in *italic font* and should provide at least one abstract method - all abstract methods are printed in *italic font*, too.

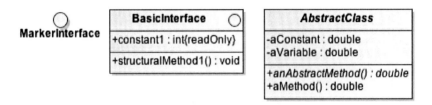

Figure B.9: UML: an interface and abstract classes

Note: *Interfaces* allow only constant attributes and public abstract Methods. Just as normal classes *Interfaces* and *Abstract Classes* can be associated and allow a higher level of structural abstraction.

B.4.2 Sequence Diagrams

To lay down how certain classes interact and to explain some algorithms a *sequence diagram* is usually employed - the sequence of the interaction is the main issue. On top the involved classes are drawn in the usual rectangular shape. Below the rectangle is a dotted line that is called *lifeline*. On the *lifeline* one class sends a message to another class - that means a method is called on the other class. The messages are drawn as lines with an arrow at the top to show the message direction. If a class will not just call another one but depending on a condition alternative methods will be called a boxed is laid over the *alternative*. See all sequence diagram parts used in that work in figure B.10.

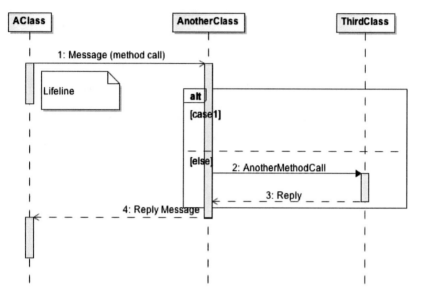

Figure B.10: UML: graphical representation of a sequence diagram

B.4.3 Composite State Diagrams

For a structural level *composite state diagrams* can be used. To summarize sets of classes or software a *component* can be used. A component might be any type software and is drawn as rectangular shape with a characteristic icon on the right side of the top compartment. A component can be *used* by another component - that shows a dependent aspect. Another way to communicate with a *component* is to use a *port*; a *port* is kind of a passageway into the *component* and out of the component. These *ports* communicate via interfaces or classes with other *components*. To emphasize that a *component* provides special interfaces these interfaces can be depicted directly from the *port* or a part of *class diagram* can be employed as a magnifier. In that case a *package* is an appropriate way of showing the coherence of the classes.

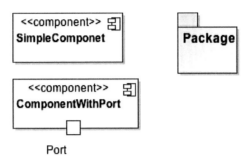

Figure B.11: UML: graphical representation of a composite state diagram

B.4.4 Summary

UML is very convenient way to show how parts of software look like and how these parts play together as a whole. The diagrams function as a magnifier that can zoom in at an arbitrary level of detail or abstraction - depending what the aim of the diagram is. During this work class diagrams are used to describe the structure of a model (see figure 5.8 on page 64) and later zooms into the generic structure (see figure 5.16 on page 81). From that level of abstraction it is quite easy to implement a database to store models.

With the algorithm described by the sequence diagram figure B.3 on page 154 one possible way is shown how to program an algorithm that deals with different types of model commands and is used as a bridge from the user interface to the database.

Using a component diagram 5.9 (p. 67) MORET as a software solution could be depicted on a very abstract level in relation to other programs.

These diagrams accompanied by the explaining text enables the user to concentrate on the abstract design much better than just one medium on its own. The sources of MORET can be downloaded on the internet for a reference implementation of a model storage engine from http://www.rosuda.org/Moret/main.html. On this web page a documentation for MORET is found as well as some hints on how to get started.

B.5 Development Comments

Many lessons were learned in course of this project, some minor and some major ones. Right from the start it was clear that models are richly structured - not just a two dimensional table of $n \times m$ observations. So the first issue was to find a structure to hold all model values. Moreover the model structure had to be preserved. Richly structured classes, specialized for one model type, proved too cumbersome. The diversity of model types today as well as new emerging types prohibit that kind of approach. The solution found was to use a generic graph structure where the edges represent the model structure. Nodes map values and additional structural information, that means the name of the structures.

Interactive model selection became possible by extracting common model structure for sets of models. That means the structured values were transformed into a two dimensional matrix. Data matrices support standard interactive tools. Even a feedback to R directly is possibly for this kind of data.

B.5.1 Recommendations

R is a very dynamic project. Accordingly not all model related data values are mapped similarly. When different models are to be compared this distinction involves extra work. Equivalent values are found at different places for related model types. Even the structural depth where these values are found is not standardized. Introducing standard names yields many benefits:

B.5.2 Unification of commands

As mentioned above variety and alterability of commands requires adjustments and extra work from the user. Unifying model commands and parameters for ease of use would solve that issue. This involves a unification of the help pages. Common parameters for related objects are presented in mixed up oreders. Example from R 2.6:

```
lm(formula, data, subset, weights, na.action,
        method = "qr", model = TRUE, x = FALSE, y = FALSE, qr = TRUE,
        singular.ok = TRUE, contrasts = NULL, offset, ...)
glm(formula, family = gaussian, data, weights, subset,
        na.action, start = NULL, etastart, mustart,
        offset, control = glm.control(...), model = TRUE,
        method = "glm.fit", x = FALSE, y = TRUE, contrasts = NULL, ...)
```

Why are subset and weights not ordered in the same way ? Using tree and name based selection MORET is able to avoid order related conflicts. Unfortunately same named values cannot be identified *automatically* when the names differ. Manual mapping is required. These problems were encountered in more exotic model types. Still neither normalization nor other means ensure stable naming, order, structure or meaning of a name.

B.5.3 Backward compatibility

The second even more important recommendation is backward compatibility. The number of R packages is growing fast. Package authors keep their package up to date and even add new function. In this update process old functions are altered sometimes. Improving software is a good idea. But if any software is modified instead of adding new functionality, old features will not work as expected. The bad part about these kinds of modification is, that old features might be used by other software. These dependencies cascade as one modified function - that is used commonly - affects every project in the dependency graph.

A way out is to write everything that is required and beyond control again. This leads to nearly identical parts of software with minor specially required features[1]. To avoid these kinds

---

[1] In software development this is in conflict with the DRY (don't repeat yourself) principle

of conflict add new functions and mark old functions to show that there is a newer and better software available. Thus old packages work properly and the dependencies don't corrupt software that would be running without these particular alterations.

B.5.4 Integration

MORET depends on R and contributed packages. But other software can benefit from data kept in MORET. One point that has been important during the development has been to manage data as well as to provide this data to other applications. Most statistic software is able to process text files. Thus MORET provides plain text formats for data exchange facilities. Additionally spread sheet tools can be fed by copy and paste. The integrated GUI supports direct data transfer to R. All these functions are provided to guarantee that favorite tools can be employed as usual. Though MORET features many integrated tools - being cooperative is even better. Cooperation integrates useful tools instead of forcing developers to rewrite their applications.

Bibliography

[Akaike 1970] AKAIKE, H.: Statistical predictor identification. In: *Annals of the Institute of Statistical Mathematics* (1970), Nr. 1, S. 203–217. http://dx.doi.org/10.1007/BF02506337. – DOI 10.1007/BF02506337

[Akaike 1973] AKAIKE, H.: Information theory and an extension of the maximumlikelihood principle. In: PETROV, B. N. (Hrsg.) ; CSAKI, F. (Hrsg.): *2nd International Symposium on Information Theory...1971*, Akademiai Kiado, 1973, 267–281

[Akaike 1974] AKAIKE, H.: A new look at the statistical model identification. In: *Automatic Control, IEEE Transactions on* 19 (1974), S. 716– 723. – ISSN 0018–9286

[Aldrich 1997] ALDRICH, J.: R. A. Fisher and the Making of Maximum Likelihood 1912-1922. In: *Statistical Science* 12 (1997), Nr. 3, 162–176. http://www.jstor.org/stable/2246367

[Allen 1974] ALLEN, D. M.: The Relationship between Variable Selection and Data Agumentation and a Method for Prediction. In: *Technometrics* 16 (1974), S. 125–127

[Anick 2004] ANICK, David: *A statistical analysis of exit polling conducted for RAW STORY.* http://www.dkosopedia.com/wiki/Academic_Papers_on_2004_Election_Results. Version: 2004

[Atkinson 1986a] ATKINSON, A. C.: Comment: Aspects of Diagnostic Regression Analysis. In: *Statistical Science* 1 (1986), Nr. 3, 379–402. http://www.jstor.org/stable/2245479

[Atkinson 1986b] ATKINSON, A. C.: *Scottish Hill Races.* http://www.statsci.org/data/general/hills.html. Version: 1986

[Atkinson et al. 2004] ATKINSON, A. C. ; RIANI, M. ; CERIOLI, A.: *Exploring Multivariate Data with the Forward Search (Springer Series in Statistics).* 1. Springer, 2004 http://amazon.com/o/ASIN/0387408525/. – ISBN 9780387408521

[Barai and Reich 1999] BARAI, S. ; REICH, Y.: Ensemble modeling or selecting the best model: Many can be better than one. In: *Artificial Intelligence for Engineering Design, Analysis and Manufacturing* 13 (1999), Nr. 5, 377-386. citeseer.ist.psu.edu/barai99ensemble.html

[Barron 1984] BARRON, A.: Predicted squared error: a criterion for automatic model selection. In: FARLOW, S. J. (Hrsg.) ; FARLOW, Stanley J. (Hrsg.) ; FARLOW, J. Farlow S. J.

(Hrsg.): *Self-Organizing Methods in Modeling: Gmdh Type Algorithms (Statistics: Textbooks and Monographs)*, Marcel Dekker Inc, 7 1984. – ISBN 9780824771614, 87–104

[Behnke and Wilmore 1974] BEHNKE, A. ; WILMORE, J. H.: *Evaluation and Regulation of Body Build and Composition (International research monograph series in physical education).* Prentice Hall, 1974 http://amazon.com/o/ASIN/0132922843/. – ISBN 9780132922845

[Berger et al. 1996] BERGER, A. ; DELLA PIETRA, S. ; DELLA PIETRA, V.: A Maximum Entropy Approach to Natural Language Processing. In: *Computational Linguistics* 22 (1996), Nr. 1, 39–71. citeseer.ist.psu.edu/article/berger96maximum.html

[Bhansali 1986] BHANSALI, R. J.: Asymptotically Efficient Selection of the Order by the Criterion Autoregressive Transfer Function. In: *Annals of Statistic* 14 (1986), Nr. 1, 315–325. http://projecteuclid.org/euclid.aos/1176349858

[Bhansali and Downham 1977] BHANSALI, R. J. ; DOWNHAM, D. Y.: Some properties of the order of an autoregressive model selected by a generalized Akaikes EPF criterion. In: *Biometrika* 64 (1977), S. 547–551

[Biernacki et al. 2000] BIERNACKI, C. ; CELEUX, G. ; GOVAERT, G.: Assessing a mixture model for clustering with the integrated completed likelihood. In: *IEEE Transactions on Pattern Analysis and Machine Intelligence* (2000), S. 719–725

[Billah et al. 2003] BILLAH, B. ; HYNDMAN, R.J. ; KOEHLER, A.B.: Empirical Information Criteria for Time Series Forecasting Model Selection / Monash University, Department of Econometrics and Business Statistics. Version: Januar 2003. http://ideas.repec.org/p/msh/ebswps/2003-2.html. 2003 (2/03). – Monash Econometrics and Business Statistics Working Papers

[Bock 1988] BOCK, H. H.: *Classification and Related Methods of Data Analysis.* North-Holland, 1988 http://amazon.com/o/ASIN/0444704043/. – ISBN 9780444704047

[Bozdogan 1987] BOZDOGAN, H.: Model selection and Akaike's information criterion (AIC): the general theory and its analytical extensions. In: *Psychometrika* 52 (1987), Nr. 3, S. 345–370. – ISSN 0033–3123

[Bozdogan 1990] BOZDOGAN, H.: On the Information-Based Measure of Covariance Complexity and its Application to the Evaluation of Multivariate Linear Models. In: *Communications in Statistics - Theory and Methods* (1990), 221–278. http://www.informaworld.com/10.1080/03610929008830199

[Bozdogan 2000] BOZDOGAN, H.: Akaike's Information Criterion and Recent Developments in Information Complexity. In: *Mathematical Psychology* 44 (2000), S. 62–91

[Breiman 1996] BREIMAN, L.: Bagging Predictors. In: *Machine Learning* 24 (1996), Nr. 2, 123–140. citeseer.ist.psu.edu/breiman96bagging.html

[Breiman and Freedman 1983] BREIMAN, L. ; FREEDMAN, D.: How many variables should be entered in a regression equation? In: *American Statistical Association* 78 (1983), S. 131–136

[Breiman et al. 1984] BREIMAN, L. ; FRIEDMAN, J. ; STONE, C. J. ; OLSHEN, R.A.: *Classification and Regression Trees.* 1. Chapman and Hall/CRC, 1984 http://amazon.com/o/ASIN/0412048418/. – ISBN 9780412048418

[Brüggemann-Klein 1993] BRÜGGEMANN-KLEIN, A.: Regular Expressions into Finite Automata. In: *Theoretical Computer Science* 120 (1993), S. 87–98

[Burnham and Anderson 2010] BURNHAM, K. P. ; ANDERSON, D. R.: *Model Selection and Multi-Model Inference: A Practical Information-Theoretic Approach.* Springer, 2010 http://amazon.com/o/ASIN/1441929738/. – ISBN 9781441929730

[Cavanaugh 1999] CAVANAUGH, J. E.: A large-sample model selection criterion based on Kullbacks symmetric divergence. In: *Statistics and Probability Letters* (1999), 333–343. http://citeseer.ist.psu.edu/cavanaugh99largesample.html

[Cavanaugh 2004] CAVANAUGH, J. E.: Criteria for linear model selection based on Kullbacks symmetric divergence. In: *Australian and New Zealand Journal of Statistics* (2004), 257–274. citeseer.ist.psu.edu/615263.html

[Chatterjee and Hadi 1986] CHATTERJEE, S. ; HADI, A. S.: Influential observations, high leverage points, and outliers in linear regression. In: *Statistical Science* (1986), Nr. 1, S. 379–416

[Chen and Bickel 2006] CHEN, A. ; BICKEL, P.: Efficient Independent Component Analysis. In: *Annals of Statistic* 34 (2006), Nr. 6, 2825–2855. citeseer.ist.psu.edu/621702.html

[Claeskens and Hjort 2003] CLAESKENS, G. ; HJORT, N. L.: The Focussed Information Criterion. In: *Journal of the American Statistical Association* 98 (2003), 900–916. http://www.math.uio.no/eprint/stat_report/2003/05-03.html

[Cook 1977] COOK, R. D.: Detection of Influential Observation in Linear Regression. In: *Technometrics* 19 (1977), Feb, Nr. 1, 15–18. http://www.jstor.org/stable/1268249

[Craven and Wahba 1979] CRAVEN, P. ; WAHBA, G.: Smoothing noisy data with spline functions: Estimating the correct degree of smoothing by the method of generalized cross-validation. In: *Numerical Mathematics* (1979), S. 377–403

[Czernichow and Muñoz 1995] CZERNICHOW, T. ; MUÑOZ, A.: Variable Selection through Statistical Sensibility Analysis: Application to Feedforward and Recurrent Neural Networks. / Swiss Federal Institute of Technology, Insituto de Investigación Tecnológica. Version: 1995. http://dx.doi.org/10.1.1.48.1091. 1995. – Forschungsbericht

[Data Description 1986] DATA DESCRIPTION, Inc.: *Data Desk*, 1986. http://www.datadesk.com/

[Davison 2003] DAVISON, A. C.: *Statistical Models (Cambridge Series in Statistical and Probabilistic Mathematics).* Cambridge University Press, 2003 http://amazon.com/o/ASIN/0521773393/. – ISBN 9780521773393

[Draper 1995] DRAPER, D.: Assessment and propagation of model uncertainty (with discussion). In: *Journal of the Royal Statistical Society* 57 (1995), 45–97. citeseer.ist.psu.edu/draper95assessment.html

[Efron 1987] EFRON, B.: *The Jackknife, the Bootstrap, and Other Resampling Plans (CBMS-NSF Regional Conference Series in Applied Mathematics).* Society for Industrial Mathematics, 1987 http://amazon.com/o/ASIN/0898711797/. – ISBN 9780898711790

[Efron and Tibshirani 1995] EFRON, B. ; TIBSHIRANI, R.: Cross-validation and the bootstrap: Estimating the error rate of a prediction rule / Standford University. Version: May 1995. `citeseer.ist.psu.edu/47726.html`. 1995 (477). – Forschungsbericht

[Efron and Tibshirani 1994] EFRON, B. ; TIBSHIRANI, R. J.: *An Introduction to the Bootstrap (Chapman & Hall/CRC Monographs on Statistics & Applied Probability)*. 1. Chapman and Hall/CRC, 1994 `http://amazon.com/o/ASIN/0412042312/`. – ISBN 9780412042317

[Feng and Liu 2003] FENG, H. ; LIU, J.: A SETAR model for Canadian GDP: non-linearities and forecast comparisons. In: *Applied Economics* 35 (2003), S. 1957–1964

[Fernandez et al. 2001] FERNANDEZ, C. ; LEY, E. ; STEEL, M.: Benchmark Priors for Bayesian Model Averaging. In: *Journal of Econometrics* 100 (2001), S. 381–427

[Foster and George 1994] FOSTER, D. P. ; GEORGE, E. I.: The Risk Inflation Criterion for Multiple Regression. In: *Annals of Statistics* 22 (1994), S. 1947–1975

[Fowler 2003] FOWLER, M.: *UML Distilled: A Brief Guide to the Standard Object Modeling Language (3rd Edition)*. 3. Addison-Wesley Professional, 2003 `http://amazon.com/o/ASIN/0321193687/`. – ISBN 9780321193681

[Freeman 2004] FREEMAN, S. F.: *The Unexplained Exit Poll Discrepancy*. `http://www.dkosopedia.com/wiki/Academic_Papers_on_2004_Election_Results`. Version: 2004

[Freund 2001] FREUND, Y.: Proceedings of the 12th Annual Conference on Computational Learning Theory. In: *An Adaptive Version of the Boost by Majority Algorithm*, Assn for Computing Machinery, 7 2001. – ISBN 9781581131673, 102–113

[Freund and Schapire 1996] FREUND, Y. ; SCHAPIRE, R. E.: Experiments with a New Boosting Algorithm. In: SAITTA, Lorenza (Hrsg.): *Machine Learning 1996 International Conference: Proceedings of the Thirteenth International Conference: Proceedings of the 13th International Conference 13th*, Morgan Kaufmann Publishers In, 7 1996. – ISBN 9781558604193, 148–156

[Friedl 2006] FRIEDL, J. E. F.: *Mastering Regular Expressions*. Third Edition. O'Reilly Media, 2006 `http://amazon.com/o/ASIN/0596528124/`. – ISBN 9780596528126

[Friedman et al. 1998] FRIEDMAN, J. ; HASTIE, T. ; TIBSHIRANI, R.: Additive logistic regression: a statistical view of boosting. In: *Annals of Statistic* 28 (1998), Nr. 2, 337–407. `citeseer.ist.psu.edu/friedman98additive.html`

[Fujikoshi and Satoh 1997] FUJIKOSHI, Y. ; SATOH, K.: Modified AIC and Cp in Multivariate Linear Regression. In: *Biometrika* (1997), S. 707–716

[Geisser 1975] GEISSER, S.: The predictive sample reuse method with application. In: *American Statistical Association* (1975), S. 320–328

[Geweke and Meese 1981] GEWEKE, J. ; MEESE, R.: Estimating regression models of finite but unknown order. In: *International Economic Review* 22 (1981), Nr. 1, S. 55–70

[Ghorbani and Owrangh 2001] GHORBANI, A. A. ; OWRANGH, K.: Stacked Generalization in Neural Networks: Generalization on Statistically Neutral Problems. In: *Neural Networks, 2001. Proceedings. IJCNN '01. International Joint Conference on* 3 (2001), 1715 – 1720. `citeseer.ist.psu.edu/494979.html`. ISBN 0–7803–7044–9

[Group 1995] GROUP, Hypersonic S.: *hsqldb - 100% Java Database*, 1995. http://hsqldb.org/

[Group 2004] GROUP, The A.: *Apache Derby*, 2004. http://db.apache.org/derby/

[Hannan and Quinn 1979] HANNAN, E. J. ; QUINN, B. G.: The determination of the order of an autoregression. In: *Royal Statistical Society* 41 (1979), S. 190–195

[Hansen and Racine 2007] HANSEN, B. E. ; RACINE, J. S.: Jackknife model averaging / University of Wisconsin and McMaster University. 2007. – Forschungsbericht

[Haring 1975] HARING, G.: *Über die Wahl der optimalen Modellordnung bei der Darstellung von stationären Zeitreihen mittels Autoregressivmodell als Basis der Analyse von EEG - Biosignalen mit Hilfe eines Digitalrechners.* 1975. – Habilitationschrift

[Hastie et al. 2003] HASTIE, T. ; TIBSHIRANI, R. ; FRIEDMAN, J. H.: *The Elements of Statistical Learning.* Corrected. Springer, 2003 http://amazon.com/o/ASIN/0387952845/. – ISBN 9780387952840

[Helbig et al. 2004] HELBIG, M. ; URBANEK, S. ; THEUS, M.: *JGR*, 2004. http://rosuda.org/JGR/index.shtml

[Ho 1998] HO, T. K.: The Random Subspace Method for Constructing Decision Forests. In: *IEEE Transactions on Pattern Analysis and Machine Intelligence* 20 (1998), Nr. 8, 832–844. citeseer.ist.psu.edu/ho98random.html

[Hoeting et al. 1999] HOETING, J. A. ; MADIGAN, D. ; RAFTERY, A. E. ; VOLINSKY, C. T.: Bayesian Model Averaging: A Tutorial. In: *Statistical Science* 14 (1999), Nr. 4, 382–417. http://www.stat.colostate.edu/~jah/papers/

[Hothorn and Buhlmann 2007] HOTHORN, T. ; BUHLMANN, P.: *mboost: Model-Based Boosting*, 2007. http://cran.r-project.org/src/contrib/Descriptions/mboost.html

[Hout et al. 2004] HOUT, Michael ; MANGELS, Laura ; CARLSON, Jennifer ; BEST, Rachel: *The Effect of Electronic Voting Machines on Change in Support for Bush in the 2004 Florida Elections.* http://www.dkosopedia.com/wiki/Academic_Papers_on_2004_Election_Results. Version: 2004

[Huang et al. 2005] HUANG, X. ; LI, S. Z. ; WANG, Y.: Jensen-Shannon boosting learning for object recognition. In: COM IEEE, Computer Society C. (Hrsg.): *2005 IEEE Computer Society Conference on Computer Vision and Pattern Recognition: Cvpr 2005, V.1-2 ...*, Institute of Electrical & Electronics Enginee, 1 2005. – ISBN 9780769523729, 144–149

[Hurvitch and Tsai 1989] HURVITCH, C. M. ; TSAI, C.-L.: Regression and time series model selection in small samples. In: *Biometrica* 76 (1989), S. 297–307

[Irizarry 2001] IRIZARRY, R. A.: Information and Posterior Probability Criteria for Model Selection in Local Likelihood Estimation. In: *Journal of the American Statistical Association* 96 (2001), Nr. 453, 303–315. citeseer.ist.psu.edu/irizarry98information.html

[Ishiguro and Sakamoto 1991] ISHIGURO, M. ; SAKAMOTO, Y.: WIC: An estimation-free information criterion. In: *Research memorandum of the Institute of Statistical Mathematics*, Institute of Statistical Mathematics, 1991, S. 410

[Ishiguro et al. 1997] ISHIGURO, M. ; SAKAMOTO., Y. ; KITAGAWA, G.: Bootstrapping log likelihood and EIC, an extension of AIC. In: *Annals of the Institute of Statistical Mathematics* (1997), S. 411–434

[Jacoby 2005] JACOBY, Bill: *Regression III.* http://polisci.msu.edu/jacoby/icpsr/regress3/. Version: 2005

[James and Hastie 1997] JAMES, G. ; HASTIE, T.: Generalizations of the Bias/Variance Decomposition for Prediction Error / Stanford University. Version: 1997. citeseer.ist.psu.edu/james97generalizations.html. 1997. – Forschungsbericht

[Jaynes 1957] JAYNES, E. T.: Information Theory and Statistical Mechanics. In: *The Physical Review* (1957), S. 620–630

[Jenkins and Watts 1968] JENKINS, G. M. ; WATTS, D. G.: *Spectral Analysis and Its Applications.* Holden Day, 1968 http://amazon.com/o/ASIN/0816244642/. – ISBN 9780816244645

[Jin et al. 2004] JIN, R. ; LIU, Y. ; SI, L. ; CARBONELL, J. ; HAUPTMANN, A. G.: A new boosting algorithm using input-dependent regularizer. In: FAWCETT, Tom (Hrsg.) ; MISHRA, Nina (Hrsg.): *Proceedings of the Twentieth International Conference on Machine Learning: August 21-24, 2003 Washington, Dc USA*, Amer Assn for Artificial, 2 2004. – ISBN 9781577351894

[Johansen 1998] JOHANSEN, T. A.: Constrained and regularized system identification. In: *Modeling, Identification and Control* 19 (1998), Nr. 2, S. 109–116. – ISSN 03327353

[Kashyap 1982] KASHYAP, R.L.: Optimal Choice of AR and MA Parts in Autoregressive Moving Average Models. In: *IEEE Transactions on Pattern Analysis and Machine Intelligence* (1982), S. 99–104

[Kehagias et al. 2001] KEHAGIAS, A. ; PETROTH, L. ; PETRIDIS, V. ; BAKIRTZIS, A. ; MASLARIS, N. ; KIARTZIS, S. ; PANAIOTOU, H.: A Bayesian multiple models combination method for time series prediction. In: *Journal of Intelligent and Robotic Systems* 31 (2001), Nr. 31, 1–3. citeseer.ist.psu.edu/petridis01bayesian.html

[Kitagawa 1996] KITAGAWA, G.: Monte Carlo Methods for the Estimation and Selection of Time Series Models. In: *Sydney International Statistical Congress, 28th Symposium on the Interface, Sydney, Australia*, 1996

[Kitagawa and Konishi 1999] KITAGAWA, G. ; KONISHI, S.: Statistical Model Evaluation by Generalized Information Criteria. In: *Bulletin of the International Statistical Institute, 52nd Session*, 1999, 485–488

[Kohavi 1995] KOHAVI, R.: A Study of Cross-Validation and Bootstrap for Accuracy Estimation and Model Selection. In: *Ijcai 1995 Conference (International Joint Conference on Artificial Intelligence//Proceedings)*, Morgan Kaufmann Publishers In, 8 1995. – ISBN 9781558603639, 1137–1145

[Lahiri 2002] LAHIRI, Parhasarathi (Hrsg.): *Model Selection (Vol: 38).* Inst of Mathematical Statistic, 2002 http://amazon.com/o/ASIN/0940600528/. – ISBN 9780940600522

[Lai et al. 2005] LAI, C. ; REINDERS, Marcel J. ; WESSELS, L.: Random Subspace Method for multivariate feature selection. In: *Pattern Recognition Letters* 27 (2005), Nr. 10

[Larsen and Hansen 1994] LARSEN, J. ; HANSEN, L. K.: Generalization Performance Of Regularized Neural Network Models. In: VLONTZOS, John (Hrsg.): *Neural Networks for Signal Processing IV: Proceedings of the 1994 IEEE Workshop : Fourth in a Series of Workshops Organized by the IEEE Signal Proce*, I.E.E.E.Press, 9 1994. – ISBN 9780780320260, 42–51

[Lebreton et al. 1992] LEBRETON, J.D. ; BURNHAM, K.P. ; CLOBERT, J. ; ANDERSON, D.R.: Modeling survival and testing biological hypotheses using marked animals. In: *Ecological Monograph* 62 (1992), S. 67–118

[Lenderman 2005] LENDERMAN, Jason: *A Note Regarding the Berkeley Survey Research Center's Paper on the Effects of Electronic Voting in Florida During the 2004 Presidential Election.* http://www.dkosopedia.com/wiki/Academic_Papers_on_2004_Election_Results. Version: 2005

[Li et al. 2003] LI, S. Z. ; SHUM, H.-Y. ; ZHANG, Z. ; ZHANG, H.J.: FloatBoost Learning for Classification. In: BECKER, Suzanna (Hrsg.) ; OBERMAYER, Klaus (Hrsg.) ; THRUN, Sebastian (Hrsg.): *Advances in Neural Information Processing Systems 15: Proceedings of the 2002 Conference (Bradford Books)*, Mit Pr, 10 2003. – ISBN 9780262025508, 993–1000

[Liu and Shum 2003] LIU, C. ; SHUM, H.-Y.: Kullback-Leibler Boosting. In: *Computer Vision and Pattern Recognition* 01 (2003), 587. http://dx.doi.org/http://doi.ieeecomputersociety.org/10.1109/CVPR.2003.1211407. – DOI http://doi.ieeecomputersociety.org/10.1109/CVPR.2003.1211407. – ISSN 1063–6919

[de Luna 1998] LUNA, X. de: An Improvement of Akaike's FPE Criterion to Reduce its Variability. In: *Journal of Time Series Analysis* (1998), S. 457–471

[Mallows 1973] MALLOWS, C.L.: Some comments on Cp. In: *Technometrics* 15 (1973), S. 661–675

[Mason et al. 2000] MASON, L. ; BAXTER, J. ; BARTLETT, P. ; FREAN, M.: Boosting Algorithms as Gradient Descent. In: LEEN, Todd K. (Hrsg.) ; MULLER, Klaus-Robert (Hrsg.) ; SOLLA, Sara A. (Hrsg.): *Advances in Neural Information Processing Systems 12: v. 12* Bd. 12, Mit Pr, 6 2000. – ISBN 9780262194501, 512–518

[McCullagh and Nelder 1989] MCCULLAGH, P. ; NELDER, J. A.: *Generalized Linear Models, Second Edition (Chapman & Hall/CRC Monographs on Statistics & Applied Probability).* 2. Chapman and Hall/CRC, 1989 http://amazon.com/o/ASIN/0412317605/. – ISBN 9780412317606

[McNaughton and Yamada 1960] MCNAUGHTON, R. ; YAMADA, H.: Regular expressions and state graphs for automata. In: *IEEE Transactions on Electronic Computers* 9 (1960), Nr. 1, S. 39–47

[McQuarrie and Tsai 1998] MCQUARRIE, A. D. R. ; TSAI, C.-L.: *Regression and Time Series Model Selection.* World Scientific Publishing Company, 1998 http://amazon.com/o/ASIN/981023242X/. – ISBN 9789810232429

[Miller 1984] MILLER, A. J.: Selection of Subsets of Regression Variables. In: *Journal of the Royal Statistical Society* 147 (1984), S. 389–425

[Moody 1994] MOODY, J.: Prediction Risk and Architecture Selection for Neural Networks. Version: 1994. `citeseer.ist.psu.edu/moody94prediction.html`. In: CHERKASSKY, V. (Hrsg.) ; FRIEDMAN, J. H. (Hrsg.) ; WECHSLER, H. (Hrsg.): *From Statistics to Neural Networks: Theory and Pattern Recognition Applications.* Springer, NATO ASI Series F, 1994, 147–165

[Moody 1991] MOODY, J. E.: Note on generalization, regularization and architecture selectionin nonlinear learning systems. In: *IEEE Workshop on Neural Networks for Signal Processing: Proceedings, 1991/91Th03855,* Inst of Electrical, 10 1991. – ISBN 9780780301184, 1–10

[Moody 1992] MOODY, J. E.: Number of Parameters: An Analysis of Generalization and Regularization in Nonlinear Learning Systems. In: *Advances in Neural Information Processing Systems* (1992), 847–854. `citeseer.ist.psu.edu/271341.html`

[Murata et al. 1994] MURATA, N. ; YOSHIZAWA, S. ; AMARI, S.-I.: Network Information Criterion—Determining the number of hidden units for an Artificial Neural Network model. In: *IEEE Transactions on Neural Networks* 5 (1994), November, Nr. 6, 865–872. `citeseer.ist.psu.edu/murata94network.html`

[Nishii 1984] NISHII, R.: Asymptotic properties of criteria for selection of variables in multiple regression. In: *Annals of Statistics* (1984), S. 758–765

[Oestereich 2001] OESTEREICH, B.: *Objektorientierte Softwareentwicklung.* Oldenbourg, 2001 `http://amazon.com/o/ASIN/3486255738/`. – ISBN 9783486255737

[P. Viola 2002] P. VIOLA, M. J.: Fast and Robust Classification using Asymmetric AdaBoost and a Detector Cascade. In: DIETTERICH, Thomas G. (Hrsg.) ; BECKER, Suzanna (Hrsg.) ; GHAHRAMANI, Zoubin (Hrsg.): *Advances in Neural Information Processing Systems 14: Proceedings of the 2001 Neural Information Processing Systems (Nips) Conference: Proceedings of ... Processing Systems (NIPS) Conference Vol 14,* Mit Pr, 9 2002. – ISBN 9780262042086, 1311–1318

[Parzen 1974] PARZEN, E.: Some recent advances in time-series modelling. In: *IEEE Transactions on Automatic Control* AC-19 (1974), S. 723–730

[Peng 2008] PENG, R. D.: *cacher,* 2008. `http://cran.r-project.org/web/packages/cacher/index.html`

[Pilone and Pitman 2005] PILONE, D. ; PITMAN, N.: *UML 2.0 in a Nutshell (In a Nutshell (O'Reilly)).* 2nd. O'Reilly Media, 2005 `http://amazon.com/o/ASIN/0596007957/`. – ISBN 9780596007959

[Pukkila and Krishnaiah 1988] PUKKILA, M.T. ; KRISHNAIAH, P.R.: On the Use of Autoregressive Order Determination Criteria in Multivariate White Noise Tests. In: *IEEE Transactions on Acoustics, Speech and Signal Processing* 36 (1988), Nr. 9, S. 1396–1403

[Quenouille 1956] QUENOUILLE, M. H.: Notes on bias in estimation. In: *Biometrika* 43 (1956), S. 353–

[Quinlan 1996] QUINLAN, J. R.: Bagging, Boosting, and C4.5. In: PRESS, AAAI (Hrsg.) ; AAAI (Hrsg.) ; ARTIFICIAL INTEL, American A. (Hrsg.): *AAAI-96: Proceedings of the Thirteenth National Conference on Artificial Intelligence and the Eighth Annual Conference on Innov:*

Proceedings of the ... Oregon (AAAI National Conference Proceedings), Mit Pr, 7 1996. – ISBN 9780262510912, 725–730

[R Development Core Team 2006] R DEVELOPMENT CORE TEAM: *R: A Language and Environment for Statistical Computing*. Vienna, Austria: R Foundation for Statistical Computing, 2006. http://www.R-project.org

[Raftery et al. 2003] RAFTERY, A. E. ; BALABDAOUI, F. ; GNEITING, T. ; POLAKOWSKI, M.: Using Bayesian Model Averaging to Calibrate Forecast Ensembles / University of Washington. Version: 2003. http://www.stat.washington.edu/www/research/reports/2003/tr440.pdf. 2003. – Forschungsbericht

[Rätsch and Warmuth 2005] RÄTSCH, G. ; WARMUTH, M. K.: Efficient Margin Maximization with Boosting. In: *Journal of Machine Learning Research* 6 (2005), 2131–2152. http://edoc.mpg.de/276539

[Rissanen 1978] RISSANEN, J.: Modeling by shortest data description. In: *Automatica* (1978), S. 465–471

[Rissanen and Ristad 1994] RISSANEN, J. ; RISTAD, E. S.: Language Acquisition in the MDL Framework. In: RISTAD, Eric S. (Hrsg.): *Language Computations: Dimacs Workshop on Human Language March 20-22, 1992 (Dimacs Series in Discrete Mathematics and Theoretical Computer Science,)*, American Mathematical Society, 9 1994. – ISBN 9780821866085, 149–166

[Schapire 1990] SCHAPIRE, R. E.: The Strength of Weak Learnability. In: *Machine Learning* 5 (1990), 197–227. citeseer.ist.psu.edu/schapire90strength.html

[Schapire 1999] SCHAPIRE, R. E.: A Brief Introduction to Boosting. In: IJCAI (Hrsg.): *Ijcai'99 2 Volume Set (International Joint Conference on Artificial Intelligence//Proceedings)*, Morgan Kaufmann Publishers In, 9 1999. – ISBN 9781558606135, 1401–1406

[Schapire 2008] SCHAPIRE, R. E.: Theoretical Views of Boosting and Applications. In: YOKOMORI, Takashi (Hrsg.) ; WATANABE, Osamu (Hrsg.): *Algorithmic Learning Theory: 10th International Conference, ALT '99 Tokyo, Japan, December 6-8, 1999 Proceedings: 10th International Workshop, ALT ... (Handbook of Experimental Pharmacology)*, Springer, 6 2008. – ISBN 9783540667483, 13–25

[Schwarz 1978] SCHWARZ, G.: Estimating the dimension of a model. In: *Annals of Statistics* 6 (1978), S. 461–464

[Seger 2006] SEGER, R.: *MORET INSTRUCTIONS*, 2006–2009. http://rosuda.org/Moret/main.html

[Sen and Srivastava 1990] SEN, A. ; SRIVASTAVA, M.: *Regression Analysis: Theory, Methods, and Applications (Springer Texts in Statistics)*. Springer, 1990 http://amazon.com/o/ASIN/0387972110/. – ISBN 9780387972114

[Shibata 1980] SHIBATA, R.: Asymptotically Efficient Selection of the Order of the Model for Estimating Parameters of a Linear Process. In: *Annals of Statistics* 8 (1980), S. 147–164

[Skilling 1998] SKILLING, J.: Massive Inference and Maximum Entropy. In: *Maximum Entropy and Bayesian Methods* 98 (1998), Nr. 1, 1-14. http://www.maxent.co.uk/documents/massinf.pdf

[Skurichina and Duin 2002] SKURICHINA, M. ; DUIN, R. P. W.: Limited Bagging, Boosting and the Random Subspace Method for Linear Classifiers. In: *Pattern Analysis And Applications* 5 (2002), 121–135. citeseer.ist.psu.edu/skurichina02limited.html

[Sohn and Dagli 2003] SOHN, S. ; DAGLI, C.H.: Combining evolving neural network classifiers using bagging. In: *International Joint Conference on Neural Networks (IJCNN),2003* Bd. 4, I.E.E.E.Press, 12 2003. – ISBN 9780780378988, 3218– 3222

[Spiegelhalter et al. 2002] SPIEGELHALTER, D. ; BEST, N. ; CARLIN, B. ; LINDE, A. van d.: Bayesian measures of model complexity and fit. In: *Royal Statistical Society* 64 (2002), 583–640. citeseer.ist.psu.edu/spiegelhalter01bayesian.html

[Stine 2004] STINE, R.A.: Model selection using information theory and the MDL principle. In: *Sociological Methods & Research* 33 (2004), S. 230–260

[Stone 1977] STONE, M.: An Asymptotic Equivalence of Choice of Model by Cross-Validation and Akaikes Criterion. In: *Journal of the Royal Statistical Society* 39 (1977), Nr. 1, S. 44–47

[Sugiyama and Ogawa 2001] SUGIYAMA, M. ; OGAWA, H.: Subspace Information Criterion for Model Selection. In: *Neural Computation* 13 (2001), Nr. 8, 1863–1889. citeseer.ist.psu.edu/sugiyama01subspace.html

[Takeuchi 1976] TAKEUCHI, K.: Distribution of information statistics and a criterion of model fitting. In: *Suri-Kagaku (Mathematical Sciences)* 153 (1976), S. 12–18

[Theus 1996] THEUS, M.: *Theorie und Anwendung Interaktiver Statistischer Graphik*. Winer-Verlag, 1996 http://amazon.de/o/ASIN/3896390511/. – ISBN 9783896390516

[Theus 2002a] THEUS, M.: Interactive Data Visualization using Mondrian. In: *Journal of Statistical Software* 7 (2002), 11, Nr. 11, 1–9. http://www.jstatsoft.org/v07/i11. – ISSN 1548-7660

[Theus 2002b] THEUS, M.: *Mondrian*, 2002. http://rosuda.org/mondrian/

[Tibshirani and Knight 1995] TIBSHIRANI, R. ; KNIGHT, K.: Model Search and Inference by Bootstrap "Bumping / University of Toronto. Version: 1995. citeseer.ist.psu.edu/tibshirani97model.html. 1995. – Forschungsbericht. – 671–686 S.

[Tibshirani and Knight 1999] TIBSHIRANI, R. ; KNIGHT, K.: The covariance inflation criterion for adaptive model selection. In: *Royal Statistical Society* B 61 (1999), 529–546. citeseer.ist.psu.edu/tibshirani99covariance.html

[Ting and Witten 1997] TING, K. M. ; WITTEN, I. H.: Stacked Generalizations: When Does It Work? In: *Proceedings of the International Joint Conference on ... (International Joint Conference on Artificial Intelligence//Proceedings)*, Elsevier Science & Technology, 8 1997. – ISBN 9781558604803, 866–873

[Ting and Witten 1999] TING, K. M. ; WITTEN, I. H.: Issues in Stacked Generalization. In: *Journal of Artificial Intelligence Research* 10 (1999), 271–289. citeseer.ist.psu.edu/ting99issues.html

[Tsuda et al. 2002] TSUDA, K. ; SUGIYAMA, M. ; MULLER, K.: Subspace information criterion for non-quadratic regularizers. In: *IEEE Transactions on Neural Networks* (2002), 70–80. citeseer.ist.psu.edu/article/tsuda02subspace.html

[Tukey 1977] TUKEY, J. W.: Exploratory Data Analysis. 1. Addison Wesley, 1977 http://amazon.com/o/ASIN/0201076160/. – ISBN 9780201076165

[Unwin et al. 2006] UNWIN, A. ; THEUS, M. ; HOFMANN, H.: Graphics of Large Datasets: Visualizing a Million (Statistics and Computing). 1. Springer, 2006 http://amazon.com/o/ASIN/0387329064/. – ISBN 9780387329062

[Urbanek 2003] URBANEK, S.: Rserve, 2003. http://www.rforge.net/Rserve/

[Urbanek 2006] URBANEK, S.: Exploratory Model Analysis: An Interactive Graphical Framework for Model Comparison and Selection. 1. Books on Demand Gmbh, 2006 http://amazon.de/o/ASIN/3833460784/. – ISBN 9783833460784

[Urbanek and Wichtrey 2003] URBANEK, S. ; WICHTREY, T.: iplots, 2003. http://cran.r-project.org/web/packages/iplots/index.html

[Vapnik and Chervonenkis 1971] VAPNIK, V. ; CHERVONENKIS, A.: On the uniform convergence of relative frequencies of events to their probabilities. In: Theory of Probability and its Applications 16 (1971), S. 264–280

[Vapnik 1998] VAPNIK, V. N.: Statistical Learning Theory. Wiley-Interscience, 1998 http://amazon.com/o/ASIN/0471030031/. – ISBN 9780471030031

[Venables and Ripley 1998] VENABLES, W. N. ; RIPLEY, B. D.: Modern Applied Statistics with S-PLUS. 2nd. Springer-Verlag Telos, 1998 http://amazon.com/o/ASIN/0387982140/. – ISBN 9780387982144

[Wallace and Dowe 1999] WALLACE, C. S. ; DOWE, D. L.: Minimum Message Length and Kolmogorov Complexity. In: The Computer Journal 42 (1999), Nr. 4, 270–283. citeseer.ist.psu.edu/wallace99minimum.html

[Wei 1992] WEI, C. Z.: On Predictive Least Squares Principles. In: The Annals of Statistics (1992), S. 1–42

[Weisberg and Wilcox 2003] WEISBERG, Herbert (Hrsg.) ; WILCOX, Clyde (Hrsg.): Models of Voting in Presidential Elections: The 2000 U.S. Election. 1. Stanford Law and Politics, 2003 http://amazon.com/o/ASIN/080474856X/. – ISBN 9780804748568

[Wilks 2005] WILKS, D. S.: Statistical Methods in the Atmospheric Sciences, Volume 100, Second Edition (International Geophysics). 2. Academic Press, 2005 http://amazon.com/o/ASIN/0127519661/. – ISBN 9780127519661

[Wolf 2008] WOLF, H. P.: relax, 2008. http://cran.r-project.org/web/packages/relax/index.html

[Wolpert 1992] WOLPERT, D. H.: Stacked Generalization / Complex Systems Group, Theoretical Division, and Center for Non-linear Studies. Version: 1992. citeseer.ist.psu.edu/wolpert92stacked.html. Los Alamos, NM : Elsevier Science Ltd., 1992 (LA-UR-90-3460). – Forschungsbericht. – 241–259 S.

[Yin and Davidson 2004] YIN, K. ; DAVIDSON, I.: Bayesian Model Averaging Across Model Spaces via Compact Encoding. In: Artificial Intelligence and Mathematics, 2004

[Zellner 1996] ZELLNER, A.: *An Introduction to Bayesian Inference in Econometrics (Wiley Classics Library)*. Wiley-Interscience, 1996 `http://amazon.com/o/ASIN/0471169374/`. – ISBN 9780471169376

[Zhao et al. 2007] ZHAO, J. ; WU, D. ; ERDOGMUS, D. ; FANG, Y. ; HE, Z.: Adaptive motion estimation schemes using maximum mutual information criterion. In: *Wireless Communications And Mobile Computing Wireless* 7 (2007), February, 205–215. `http://dx.doi.org/10.1002/wcm.v7:2`. – DOI 10.1002/wcm.v7:2. – ISSN 1530–8669

List of Figures

List of Tables

List of Algorithms

Index

Ralf Seger

has been working as a professional programmer and as an IT consultant in Munich. He received a Ph. D. from Augsburg University, the home of MORET. His research interest is to improve current software and methods. Another aim is to create better, user-friendly, software so that a broader audience benefits from many tools that seem incomprehensible now.